HAESE MATHEMATICS

Mathematics

for the international student
Mathematical Studies SL

third edition

Mal Coad
Glen Whiffen
Sandra Haese
Michael Haese
Mark Humphries

for use with
IB Diploma Programme

Mal Coad
Glen Whiffen
Robert Haese
Michael Haese

third edition

 Haese Mathematics

EXAM PREPARATION & PRACTICE GUIDE

EXAM PREPARATION & PRACTICE GUIDE third edition

Mal Coad   B.Ec, Dip.T.
Glen Whiffen   B.Sc., B.Ed.
Robert Haese   B.Sc.
Michael Haese   B.Sc.(Hons), Ph.D.

Haese Mathematics
152 Richmond Road, Marleston, SA 5033, AUSTRALIA
Telephone: +61 8 8210 4666, Fax: + 61 8 8354 1238
Email: info@haesemathematics.com.au
Web: www.haesemathematics.com.au

National Library of Australia Card Number & ISBN 978-1-921972-07-2

© Haese & Harris Publications 2013

Published by Haese Mathematics
152 Richmond Road, Marleston, SA 5033, AUSTRALIA

First Edition	2006
Second Edition	2010
Third Edition	2013
Reprinted	2015

Artwork by Gregory Olesinski, Brian Houston and Piotr Poturaj.
Cover design by Piotr Poturaj.

Typeset in Australia by Charlotte Frost and Deanne Gallasch.

Typeset in Times Roman 9/10.

Printed in China by Prolong Press Limited.

The Guide has been developed independently of the International Baccalaureate Organization (IBO). The Guide is in no way connected with, or endorsed by, the IBO.

Acknowledgements: While every attempt has been made to trace and acknowledge copyright, the authors and publishers apologise for any accidental infringement where copyright has proved untraceable. They would be pleased to come to a suitable agreement with the rightful owner.

The authors and publishers would like to thank all those teachers who have read the proofs of this book and offered advice and encouragement.

FOREWORD

The aim of this Guide is to help you prepare for the Mathematical Studies SL final examinations.

The Guide covers Topics 1-7 in the syllabus. Each topic begins with a concise summary of key facts. This is intended to complement your textbook. Following each summary is a set of "short questions" and "long questions" which can be used to consolidate your knowledge by reminding you of the fundamental skills required for the topic.

The best way to consolidate your mathematical understanding is by being active. This includes summarising the topics in an effective way and attempting questions.

This Guide also offers five specimen examinations with each examination divided into Paper 1 and Paper 2 questions. Each Paper 1 has fifteen questions to be completed within 90 minutes; each Paper 2 has six questions also to be completed within 90 minutes. The specimen examinations reflect a balanced assessment of the course.

Fully worked solutions are provided for every question in the topic section of the Guide. Detailed marks schemes are provided for each paper in the specimen examinations. In many situations alternative solutions exist which are accepted.

Try to complete the specimen examinations under examination conditions. Getting into good habits will reduce pressure during the examination.

- It is important that you persevere with a question, but sometimes it is a good strategy to move on to other questions and return later to ones you have found challenging. Time management is very important during the examination and too much time spent on a difficult question may mean that you do not leave yourself sufficient time to complete other questions.

- Use a pen rather than a pencil, except for graphs and diagrams.

- If you make a mistake draw a single line through the work you want to replace. Do not cross out work until you have replaced it with something you consider better.

- Set out your work clearly with full explanations. Do not take shortcuts.

- Diagrams and graphs should be sufficiently large, well labelled and clearly drawn.

- Remember to leave answers correct to three significant figures unless an exact answer is more appropriate or a different level of accuracy is requested in the question. Answers involving money can be given to two decimal places, three significant figures, or to the nearest integer.

Get used to reading the questions carefully.

- Check for key words. If the word "hence" appears, then you must use the result you have just obtained. "Hence, or otherwise" means that you can use any method you like, although it is likely that the best method uses the previous result.

- Rushing into a question may mean that you miss subtle points. Underlining key words may help.

- Often questions in the examination are set so that, even if you cannot get through one part, the question can still be picked up in a later part.

After completing a practice set, identify areas of weakness.

- Return to your notes or textbook and review the Topic.

- Ask your teacher or a friend for help if further explanation is needed.

- Summarise each Topic. Summaries that you make yourself are the most valuable.

- Test yourself, or work with someone else to help improve your knowledge of a Topic.

- If you have had difficulty with a question, try it again later. Do not just assume that you know how to do it once you have read the solution. It is important that you work on areas of weakness, but do not neglect the other areas.

Your graphics display calculator is an essential aid.

- Make sure you are familiar with the model you will be using.

- In trigonometry questions remember to check that your graphics calculator is in degrees.

- Become familiar with common error messages and how to respond to them.

- Important features of graphs may be revealed by zooming in or out.

- Asymptotic behaviour is not always clear on a graphics calculator screen; don't just rely on appearances. As with all aspects of the graphics calculator, reflect on the reasonableness of the results.

- Are your batteries fresh?

We hope this guide will help you structure your revision program effectively. Remember that good examination techniques will come from good examination preparation.

We welcome your feedback:

web: http://haesemathematics.com.au

email: info@haesemathematics.com.au

TABLE OF CONTENTS

MATHEMATICAL LANGUAGE

You should understand these mathematical terms, and be able to use them appropriately:

- addition, subtraction, multiplication, division
- sum, difference, product, quotient
- index or exponent, base
- prime, composite
- factors, prime factors, common factors, highest common factor (HCF)
- multiples, lowest common multiple (LCM)

NUMBER SETS

You should recognise these special number sets:

- natural numbers $\mathbb{N} = \{0, 1, 2, 3, 4, ...\}$
- integers $\mathbb{Z} = \{...., -2, -1, 0, 1, 2, ...\}$
- rational numbers $\mathbb{Q} = \left\{ \dfrac{p}{q} \mid p, q \in \mathbb{Z}, \ q \neq 0 \right\}$
- real numbers $\mathbb{R} = \{$all numbers on the number line$\}$

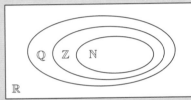

All terminating and recurring decimal numbers are rational.

THE FUNDAMENTAL THEOREM OF ARITHMETIC

Every composite number can be written as the product of prime factors in exactly one way (ignoring order).

ORDER OF OPERATIONS

Brackets

Exponents

Divisions
Multiplications } as you come to them

Additions
Subtractions } as you come to them

SCIENTIFIC NOTATION (STANDARD FORM)

$$a \times 10^k \quad \text{where} \quad 1 \leqslant a < 10 \quad \text{and} \quad k \in \mathbb{Z}$$

For example:
- 2565 in scientific notation is 2.565×10^3
- 3.04×10^{-3} as a decimal is $0.003\,04$

INTERNATIONAL SYSTEM OF UNITS (SI UNITS)

You should understand and be able to use SI units, in particular those for distance, mass, temperature, and time.

You should be able to:

- perform metric conversions $\quad 5.3 \text{ kg} = 5300 \text{ g}$
- use areas and volumes $\quad 8 \text{ m} \times 5 \text{ m} = 40 \text{ m}^2$
  $3 \text{ m} \times 2 \text{ m} \times 2 \text{ m} = 12 \text{ m}^3$
- use formulae to convert between temperature units
- use rates where we divide units

$$\text{speed (m s}^{-1}) = \frac{\text{distance (m)}}{\text{time (s)}}$$

APPROXIMATION AND ESTIMATION

Rounding numbers:

- If the digit after the one being rounded is less than 5, round down.
- If the digit after the one being rounded is 5 or more, round up.

Significant figures are counted from the first non-zero digit from the left.

For example:

$3.413 \approx 3.41$	(to 2 decimal places)
$0.034\,56 \approx 0.0346$	(to 3 significant figures)
$236.5 \approx 237$	(to the nearest whole number)

You are expected to give answers either exactly or rounded to 3 significant figures unless otherwise specified in the question.

A **measurement** is accurate to $\pm\frac{1}{2}$ of the smallest division on the scale.

An **approximation** is a value given to a number which is close to, but not equal to, its true value.

An **estimation** is a value which is found by judgement or prediction instead of carrying out a more accurate measurement.

If the exact value is V_E and the approximate value is V_A then:

- **error** $= V_A - V_E$
- **percentage error** $E = \dfrac{|V_A - V_E|}{V_E} \times 100\%$

CURRENCY CONVERSION

For simple currency conversions where there is **no commission**, we simply multiply or divide by the **currency exchange rate**.

You should be able to perform currency conversions from data given in tables.

For example:

		Currency converting to		
		AUD	GBP	USD
Currency converting from	AUD	1	0.6	0.8
	GBP	1.6	1	1.33
	USD	1.25	0.75	1

$$1 \text{ AUD} = 0.8 \text{ USD}$$
$$\therefore \ 500 \text{ AUD} = 500 \times 0.8 \text{ USD}$$
$$= 400 \text{ USD}$$

Commission on currency exchange is either:

- a fixed percentage of the transaction, or
- the difference between buying and selling rates.

If buying and selling rates are quoted, they are relative to the broker.

COMPOUND INTEREST

For interest compounding k times per year,

$$FV = PV \times \left(1 + \frac{r}{100k}\right)^{kn}$$

where FV is the future value or final balance
PV is the present value or amount originally invested
r is the interest rate per year
n is the number of years.

The interest earned $= FV - PV$

You should be able to use the TVM solver on your calculator to solve compound interest problems.

DEPRECIATION

Depreciation is the loss in value of an item over time. It is assumed to compound annually.

The depreciation formula is $FV = PV \times \left(1 + \frac{r}{100}\right)^{n}$

where FV is the future value after n time periods
PV is the original purchase price
r is the depreciation rate per period and r is negative
n is the number of periods.

LAWS OF ALGEBRA

Index or Exponent Laws	
$a^m \times a^n = a^{m+n}$	$\left(\frac{a}{b}\right)^n = \frac{a^n}{b^n}, \ b \neq 0$
$\frac{a^m}{a^n} = a^{m-n}, \ a \neq 0$	$a^0 = 1, \ a \neq 0$
$(a^m)^n = a^{m \times n}$	$a^{-x} = \frac{1}{a^x}$ and $\frac{1}{a^{-x}} = a^x$
$(ab)^n = a^n b^n$	

Expansion and Factorisation Laws	
Distributive law	$a(b+c) = ab + ac$
FOIL rule	$(a+b)(c+d) = ac + ad + bc + bd$
Difference of two squares	$(a+b)(a-b) = a^2 - b^2$
Perfect squares	$(a+b)^2 = a^2 + 2ab + b^2$ $(a-b)^2 = a^2 - 2ab + b^2$

EQUATIONS AND FORMULAE

You should understand the meanings of: expression, equation, inequality, formula, substitution, evaluate.

Linear equations can be written in the form $ax + b = 0$ where x is the **variable** and a, b are constants.

You should be able to solve linear equations:

- using inverse operations
- using technology.

You should be able to **rearrange** formulae to make another variable the **subject**.

Linear simultaneous equations occur when there are two equations in two unknowns. You should be able to use technology to solve linear simultaneous equations.

QUADRATIC EQUATIONS

Quadratic equations have the form $ax^2 + bx + c = 0, \ a \neq 0$.

The solutions of the equation are called its **roots**.

These values for x are also called the **zeros** of the quadratic expression $ax^2 + bx + c$.

If $x^2 = k$ then $\begin{cases} x = \pm\sqrt{k} & \text{if } k > 0 \\ x = 0 & \text{if } k = 0 \\ \text{there are no real solutions if } k < 0. \end{cases}$

If a quadratic equation is given in factorised form, solve using the **Null Factor law**.

For other quadratic equations, solve using technology.

SEQUENCES

A **number sequence** is a list of numbers which follow a pattern.

The numbers in a sequence are called its **members** or **terms**.

We can define a sequence by:

- listing terms
- using words
- a formula for the **general term** or **nth term**.

ARITHMETIC SEQUENCES

An **arithmetic sequence** is a sequence in which each term differs from the previous one by the same fixed number.

$\{u_n\}$ is arithmetic $\Leftrightarrow \ u_{n+1} - u_n = d$ for all $n \in \mathbb{Z}^+$, where d is a constant called the **common difference**.

The general term of an arithmetic sequence is $u_n = u_1 + (n-1)d$.

An application of arithmetic sequences is **simple interest**.

GEOMETRIC SEQUENCES

A sequence is **geometric** if each term can be obtained from the previous one by multiplying by the same non-zero constant.

$\{u_n\}$ is geometric $\Leftrightarrow \ \frac{u_{n+1}}{u_n} = r$ for all $n \in \mathbb{Z}^+$,

where r is a constant called the **common ratio**.

The general term of a geometric sequence is $u_n = u_1 r^{n-1}$.

Applications of geometric sequences include **compound interest**, and **growth** and **decay** problems.

SERIES

A **series** is the addition of the terms of a sequence.

For the sequence $\{u_n\}$, the series including the first n terms is $u_1 + u_2 + u_3 + ... + u_n$.

The result of this addition is $S_n = u_1 + u_2 + u_3 + ... + u_n$.

For an **arithmetic series**:

$$S_n = \frac{n}{2}(u_1 + u_n) \quad \text{or} \quad S_n = \frac{n}{2}(2u_1 + (n-1)d)$$

For a **geometric series**:

$$S_n = \frac{u_1(r^n - 1)}{r - 1} \quad \text{or} \quad S_n = \frac{u_1(1 - r^n)}{1 - r}$$

SKILL BUILDER - SHORT QUESTIONS

1 Place the following numbers in the appropriate region of the Venn diagram.

$2.5, \pi, -3, \frac{8}{5}, 0, \sqrt{25}$

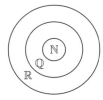

2 **a** Draw a number line to represent the set
$X = \{x \mid -3 \leqslant x \leqslant 3, \ x \in \mathbb{R}\}$

   b Use a number line to clearly represent each of the following:
   i $x < -1$ **ii** $0 \leqslant x < 2$ **iii** $x \geqslant 2$

3 If $a = 2.5$, $b = 7$, and $c = 137$, calculate $\dfrac{a^2 + b}{c}$.

Give your answer:

   a to 2 decimal places

   b to 3 significant figures

   c to 3 significant figures in scientific notation.

4 Gordon needs to fertilise the large garden landscape shown in the diagram:

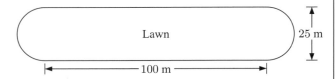

The ends of this garden landscape are semi-circular, and the middle is rectangular.

Gordon calculates the area of the garden using correct formulae, but approximates the value of π to one significant figure, so $\pi \approx 3$.

   a Write down Gordon's approximation for the area of the garden landscape.

   b Calculate the percentage error in Gordon's approximation. Round your answer to 2 significant figures.

5 In the table below, indicate using a Y (yes) or N (no) which number sets the given numbers are members of:

	$\mathbb{N}$	$\mathbb{Z}$	$\mathbb{Q}$	$\mathbb{R}$
$(-2)^3$				Y
$\sqrt{2}$	N			
0.65		N		
1.27×10^4			Y	

6 Evaluate $\sqrt{\dfrac{32.76}{3.95 \times 2.63}}$, giving your answer:

   a correct to 3 decimal places

   b correct to the nearest whole number

   c correct to 3 significant figures

   d in standard form.

7 The speed of light is approximately $186\,280$ miles per second. The distance from Mars to Earth is approximately $195\,000\,000$ km.

   a Given that mile : kilometre $= 1 : 1.609$, determine the speed of light in kilometres per minute. Give your answer to 3 significant figures.

   b Express your answer to **a** using scientific notation.

   c If a light source on Mars is ignited, how many minutes will it be before it can be seen through a telescope on Earth?

8 Consider the arithmetic sequence 120, a, 98, b, ...

   a Determine the values of a and b.

   b Write down the nth term of the sequence.

   c Determine the number of positive terms in this sequence.

9 Lisa invests £18 000 into a savings account which pays 4.5% p.a. interest, compounded annually.

   a Calculate the value of Lisa's investment after:
   i 1 year **ii** 2 years.

   b Write down a formula which calculates the value of the investment after n years.

   c Calculate the value of the investment after 13 years.

   d Determine the minimum length of time required for the investment to be triple its original value.

10 **a** A measurement of 5.645 cm is rounded to 3 significant figures.
   i Write down the actual error caused by rounding.
   ii Calculate the percentage error.

   b The speedometer of a car reads 70 km h^{-1}. It is accurate to within 3.2%.
   i What is the maximum possible error?
   ii Write down the extreme possible values for the true speed of the car.

11 The first three terms of a sequence are -2, -9, and -16.

   a Write down the next two terms of the sequence.

   b Find a formula for the nth term of the sequence.

12 The exchange rate from United States Dollars (USD) to British Pounds (GBP) is $1 : 0.6439$.

   a Determine the exchange rate from GBP to USD.

   b How many British pounds will I receive when I exchange USD 365?

   c How much USD would I receive for GBP 500?

13 A bank's exchange rate for 1 euro is 125.3 Japanese yen. The bank charges 1% commission on each transaction.

   a How many yen would be received if €1000 was exchanged?

   b How many euros would be received if 46 000 yen was exchanged?

14 A sequence is defined by $u_n = n(n + 1)$.

   a Write down the first 3 terms of the sequence.

   b Find the 15th term.

   c Which term of this sequence is 600?

15 Copy and complete the table by placing ticks in the appropriate boxes.

	$\sqrt{4}$	-2	3.75	π	$2.\overline{3}$
$\mathbb{N}$					
$\mathbb{Z}$					
$\mathbb{Q}$					
$\mathbb{R}$					

16 Find the value of $2.34 + \dfrac{5.25}{3.10 \times 7.65}$, giving your answer:

  a correct to the nearest integer

  b correct to 4 decimal places

  c correct to 3 significant figures

  d in scientific notation.

17 The first three terms of an arithmetic sequence are -347, $k - 166$, and -185.

  a Find the value of k.

  b Find a formula for the nth term of the sequence.

  c Which is the first positive term of the sequence?

18 In order to save for a holiday to Jamaica, Yasmin invests €2500 for 2 years.

  a Calculate the final value of her investment if she receives an interest rate of 5.5% p.a. compounded quarterly over this time period.

  b Prior to leaving for Jamaica, Yasmin converts her final investment amount into US dollars (USD). Given that 1 USD is exchanged for 0.7437 euros, calculate the amount that Yasmin will receive.

19 £14 000 is invested at 6% p.a. compounded monthly.

  a Calculate the total value of the investment after 5 years, giving your answer correct to 2 decimal places.

  b Calculate the minimum whole number of months required for this investment to double in value.

20 A bank in Scotland offers the following rates for changing 1 British pound (GBP) to Canadian dollars (CAD): 'buy at 1.589', 'sell at 1.513'.

  a Convert GBP 300 to CAD.

  b Convert CAD 300 to GBP.

21 The sixth term of an arithmetic sequence is 49 and the fifteenth term is 130.

  a Find the common difference for this sequence.

  b Find the first term.

  c How many terms of this sequence have a value less than 300?

22 A section of tubing is formed into a circle with diameter 4.5 m.

  a Calculate the length of the tubing.

  b Round your answer to the nearest whole number.

  c Calculate the percentage error when rounding the answer to the nearest whole number.

23 The first three terms of a geometric sequence are 0.75, 2.25, and 6.75.

  a Find the common ratio.

  b Write down a formula for the nth term.

  c Calculate the sum of the first 10 terms.

24 The sum of the first 7 terms of an arithmetic series is 329. The common difference is 14.

  a Find the value of the first term.

  b Find n given that the sum of the first n terms of the sequence is 69 800.

25 A currency exchange service changes US dollars (USD) to Swiss francs (CHF) using: 'buy at 0.9384', 'sell at 0.8917'. Shelley wants to exchange 400 USD to Swiss francs.

  a How much will she receive?

  b If Shelley changes the francs back to US dollars immediately, how many dollars will she get?

  c How much has the exchange of currencies cost Shelley?

26 **a** Determine the value after 4 years and 3 months if £4750 is invested at a nominal rate of 6.7%, compounded monthly.

  b How much interest has been earned?

  c How much less interest would have been earned if the interest had been compounded annually?

27 One stage of a cycling race covers 80 km on the flat, and the remainder is in the mountains. The cyclists can average 45 km h^{-1} on the flat and 25 km h^{-1} in the mountains.

  a Determine the time taken by the cyclists to cover the flat section of the course. Give your answer correct to the nearest minute.

  b If the total time taken for the stage is 3 hours, find the distance covered in the mountains.

28 Consider the arithmetic sequence ...., 27, x, 42, ... where 27 is the fifth term of the sequence.

  a Calculate the common difference.

  b Find the first term of the sequence.

  c Find the first term of the sequence which is greater than 2400.

29 The sequence 45, x, 281.25, ... is geometric.

  a Find the common ratio r given that $x > 0$.

  b Find the sum of the terms $u_5 + u_6 + ... + u_{12}$.

30 The length of a section of pipe is stated as 4 m. Claudia carefully measures the pipe and finds the actual length to be 3.94 m.

  a Write down the size of the error in the stated length.

  b Five such sections of pipe are joined together. Find the actual length of the joined pipes.

  c Write down the error for the joined pipes.

  d Calculate the percentage error of the stated length against the actual length of the joined pipes.

31 A rubber ball is dropped from a height of 5 m. It bounces up to a height of 4.5 m on the first bounce, then to 4.05 m on the second bounce, and so on.

  a Find the common ratio of the sequence formed by these numbers.

  b Calculate the height of the third bounce.

  c How far has the ball travelled vertically by the time it strikes the floor for the fourth time?

32 **a** List the elements of the following sets:

   i $A = \{x \mid -2 < x < 3, \ x \in \mathbb{Z}\}$

   ii $B = \{\text{prime numbers less than } 15\}$

   iii $C = \{x \mid x^2 = 8, \ x \in \mathbb{R}\}$

  b State whether the following statements are true or false:

   i All rational numbers are integers.

   ii $4 - p > 4 + p, \ p \in \mathbb{Z}^-$

   iii $\mathbb{N} = \{0, 1, 2, 3, ...\}$

33 The exchange rate from Japanese yen (JPY) to Australian dollars (AUD) is $1 : 0.010 334$.
A commission of 1.5% is charged per transaction.

  a Determine the commission in JPY charged for changing 60 000 yen to Australian dollars.

b How many dollars would be received?

c If the dollars were converted back to yen immediately, how many yen would there be?

34 Joanne recently won €40 000 in a lottery and decided to invest it for 5 years.

  a Determine which of the following terms would generate the greater interest on Joanne's winnings:
- 5.05% p.a. compounded monthly
- 5.20% p.a. compounded yearly.

  b Calculate the extra interest earned if she chooses the best terms.

35 The value of a car decreases by 12.5% each year. After 3 years its value is $21 400.

  a Find the original value of the car, giving your answer correct to the nearest $100.

  b In which year will the value of the car fall below $10 000?

36 128 football clubs enter the first round of a knockout competition. In each round, a half of the participants are eliminated.

  a How many clubs participate in the:

    i second round   **ii** third round?

  b If there are n rounds, how many participants remain in the nth round?

  c Calculate the number of rounds needed to determine a winner.

37 The cost of 15 books and 8 pens is 209 NZD. The cost of 7 books and 3 pens is 96.25 NZD.

  a Write two equations involving b and p which represent this information.

  b Solve the equations simultaneously to find b and p.

  c Find the cost of 10 books and 6 pens.

38 The second term of a geometric sequence is 14.5 and the fifth term is 1.8125.

  a Determine the common ratio.

  b Find the value of the first term.

  c Find the sum of the first 5 terms.

39 The population of a small town increases by an average of 9% per annum. In 2010, the population was 1200.

  a Calculate the size of the population in 2015.

  b In which year will the population reach 2500?

  c Find the rate of increase that would result in the population reaching 3200 in 2020.

40 Consider the sequence $2k - 9$, k, $k - 6$, ...

  a Determine the two possible values of k for which this sequence is geometric.

  b For what value of k is the sequence arithmetic?

41 **a** Solve, using technology:

    i $12s + 17r = 277$ **ii** $u_1 + 27d = 162$
       $5s + 11r = 135$ $35d = 202 - u_1$

  b The cost of hiring a taxi includes a flat fee of $\$a$ plus $\$p$ per km. A 12 km taxi ride costs $20, and a 22 km journey costs $34. Find the values of a and p.

42 Some furniture was purchased for €6000. After 5 years its value has depreciated to €1200.

  a Calculate the average annual rate of depreciation.

  b List the annual book value at the end of each year, assuming the furniture did depreciate by the rate found in **a**.

43 **a** How long will it take for $6500 to earn at least $2000 interest at a nominal rate of 8%, compounded quarterly?

  b Calculate the rate of interest compounded monthly, which would earn the same amount of interest over the same period of time.

44 **a** Find the term that the sequences $u_n = 178 - 4n$ and $u_n = 7n + 57$ have in common.

  b A firm's revenue function is $R = 25n$ and its cost function is $C = 21\,000 + 7.5n$, where n is the number of goods produced and sold. Find the value of n such that the cost equals the revenue.

  c The sum of the first n natural numbers is equal to $\dfrac{n(n+1)}{2}$. For what values of n does the sum exceed 435?

45 The nth term of a geometric sequence is given by

$u_n = 4374 \left(\frac{1}{3}\right)^{n-1}$, where n is a positive integer.

  a Determine the value of the 7th term of the sequence.

  b For what values of n is $u_n \leqslant 0.1$?

  c Determine the sum of the first 8 terms of the sequence.

46 The perimeter of a rectangle is 80 cm. The width is x cm.

  a Write down the value of the length, in terms of x.

  b Show that the area of the rectangle is given by the function $A = 40x - x^2$ cm^2.

  c If the area of the rectangle is 375 cm^2, find its dimensions.

47 **a** A computer was purchased for 4500 CAD. It depreciated in value at 22.5% p.a. Find its book value after 3 years.

  b Five years after its purchase, a tractor has book value £29 000. It depreciated at 17.5% p.a. during this time.

    i Find the original cost of the tractor, to the nearest pound.

    ii Determine the total amount of depreciation for the 5 years.

    iii Suppose the tractor had been leased for the 5 years at a fixed amount per year. How much would this fixed amount need to be, for the total value of the lease to match the loss due to depreciation?

48 The height of a sky rocket above the ground t seconds after firing is given by $s = ut - 5t^2$ m, where u represents the initial speed of the sky rocket. If the initial speed is 70 m s^{-1}, find:

  a the amount of time the rocket is in the air

  b the time the rocket is above 30 m.

49 Karen purchases a new mobile phone. The monthly cost $\$C$ of the phone when used for t minutes per month is $C = rt + b$, where b is a fixed monthly fee, and r is the call cost for each minute of use. In the first month, Karen paid $19.15 for 23 minutes of calls. In the second month, she paid $15.95 for 15 minutes of calls. Find the values of b and r.

SKILL BUILDER - LONG QUESTIONS

1 **a** The nth term of an arithmetic sequence is given by $u_n = 4 + 11n$. Find:
    i the first 2 terms of the sequence
    ii the sum of the first 10 terms.

    b The nth term of a geometric sequence is given by $u_n = 4(2.2)^{n-1}$. Find:
    i the fifth term of the sequence
    ii the sum of the first 10 terms.

    c Find the first term for which the value of the geometric sequence is greater than the value of the arithmetic sequence. What is the difference between the terms of the two sequences for this value of n?

2 **a** Draw a number line to represent $x \in \mathbb{R}$, and mark the integers from -5 to 5. Represent the following on the number line using appropriate notation.
    i $x \geqslant 4$ **ii** $1 \leqslant x < 3$
    iii $-3 \leqslant x \leqslant 0$ **iv** $x < -4$

    b Let $X = \{x \mid -4 \leqslant x \leqslant 4, \ x \in \mathbb{Z}\}$ and $Y = \{y \mid -4 \leqslant y \leqslant 4, \ y \in \mathbb{Z}\}$.
On a set of coordinate axes, plot the points which represent the set:
    i $W = \{(x, y) \mid \frac{y}{x} = -1, \ x \in X, \ y \in Y\}$
    ii $V = \{(x, y) \mid x \geqslant 0, \ y \geqslant 0, \ y - x = 1, \ x \in X, \ y \in Y\}$

    c If $x, y \in \mathbb{Q}$ and $x > y$, write down an example which makes $x^2 < y^2$ true.

3 The fifth term of an arithmetic sequence is 51, and the sum of the first 5 terms is 185.

    a If u_1 is the first term of the sequence, and d is the common difference, write down two equations in u_1 and d which satisfy the information provided.

    b Solve the equations from **a** to determine the values of u_1 and d.

    c Find the first term in this sequence to exceed 1000.

    d The sum of the first k terms in this sequence is 3735.
    i Deduce that k satisfies the equation $7k^2 + 39k - 7470 = 0$.
    ii Hence, find k.

4 **a** Which of the following statements are false? Justify your answers.
    i $\{-2, -1, 0, 1, 2\} \subset \{x \mid x < 2, \ x \in \mathbb{R}\}$
    ii $\{0, 1, 2, 3, 4\} \subset \{x \mid x \leqslant 5, \ x \in \mathbb{Z}\}$
    iii $\{x \mid x^2 + x = 2, \ x \in \mathbb{Z}\} = \{-1, 1, 2\}$

    b Let $U = \{x \mid 14 \leqslant x < 30, \ x \in \mathbb{N}\}$,
    $A = \{$multiples of $7\}$, $B = \{$factors of $56\}$,
    $C = \{$even numbers $\geqslant 20\}$.
    i List the members of each of the sets A, B, and C contained in U.
    ii Represent these sets on a 3-circle Venn diagram.

    c p and q are different integers. Which of the following statements is false? Give an example to support your decision.
    A $p + q = q + p$ **B** $p - q = q - p$ **C** $pq = qp$

5 **a** Evaluate $25.32 \times \dfrac{6.057}{2.4 \times \sqrt{5.14}}$, giving your answer correct to:
    i five significant figures   **ii** the nearest tenth
    iii one significant figure.

    b Three sections of fencing are erected. Each section has a stated length of 3.60 m, measured to the nearest tenth of a metre. The actual length of each section is 3.63 m.
    i Find the actual length covered by the three sections of fencing.
    ii Calculate the percentage error between the actual length and the stated length of the three sections of fencing.

    c The three sections of fencing in **b** form one side of a square enclosure. The enclosure will have a concrete floor 100 ± 5 mm thick. Concrete costs €47.50 per cubic metre.
    i Write down the minimum and maximum possible values for the volume of concrete needed for the floor.
    ii Calculate the difference in cost between the maximum possible volume of concrete and the planned volume based on the stated length.
    iii Express the cost difference in **ii** as a percentage of the planned cost.

6 **a** Write down the first three terms of the following sequences:
    i $u_n = 120 + 3(n - 1)$
    ii $u_{n+1} = u_n + 7, \ u_1 = 4$.

    b Which term do the sequences in **a** have in common, and what is the value of that term?

    c Which of the sequences in **a** has 151 as one of its terms?

    d The sum of the first n terms of the sequences in **a** is the same. Find n.

    e When will the sum of the first sequence in **a** exceed the sum of the second sequence by 228?

7 Dean travels from Switzerland to the United Kingdom, then the United States of America. The table below contains the currency conversion information relevant for his trip.

	CHF	GBP	USD
CHF	1.000	0.700	q
GBP	p	1.000	1.550
USD	1.085	0.645	r

    a Calculate the values of p, q, and r, giving answers correct to 3 decimal places.

    b In Switzerland, Dean converts $20\,000$ Swiss francs (CHF) into British pounds (GBP). How much money will he receive?

    c Upon leaving the United Kingdom, Dean converts his remaining 8000 British pounds into US dollars (USD). How much money will he receive?

    d Dean returns to Switzerland with 2500 US dollars. He converts this back into Swiss francs, but is charged 2% commission on this transaction. How much money will Dean receive, to the nearest franc?

    e In preparation for his next holiday, Dean invests all his remaining Swiss francs into a savings account at 5.1% p.a. interest compounded monthly. How many whole months must pass before Dean's investment is doubled?

8 A pool has two rectangular sections, the swimming pool that is 180 cm deep, and the wading pool that is 50 cm deep. The dimensions of these regions are shown in the diagram below.

swimming pool 180 cm deep	wading pool 50 cm deep	$(x-2)$ m
$(x+8)$ m	x m	

a Show that the total area of the surface of the pool is given by the expression $2x^2 + 4x - 16$.

b Given that the total surface area of the pool is 224 m^2, find the value of x.

c Calculate the volume of water (in litres) that the pool contains.

d Over a period of time, 2400 litres of water evaporates from the wading pool.

   i Find the volume of water, in litres, that the wading pool now contains.

   ii Find the new depth of the wading pool.

9 In the compound interest formula

$$FV = PV\left(1 + \tfrac{r}{100k}\right)^{kn}$$

the term $\left(1 + \tfrac{r}{100k}\right)$ is often called the *multiplier*.

The table below shows the value of the multiplier for various interest rates and compounding periods.

Compound period	6%	7.5%	10%	12.5%
annual	1.06	1.075	1.10	c
quarterly	1.015	a	1.025	1.031 25
monthly	1.005	1.006 25	b	1.010 417

a Find the values of a, b, and c.

b Using the values from the table:

   i show that an investment of 4000 CHF will grow to 5813.18 CHF over 5 years at a nominal rate of 7.5% p.a. compounded monthly

   ii find how long it will take for this investment to exceed 5000 CHF at 10% p.a. compounded quarterly.

c Determine the amount needed to be invested now, at 12.5% p.a. compounded monthly, for the future value of the investment to be 10 000 CHF after 7 years.

d The exchange rate from USD to Swiss francs (CHF) is 1:1.2778, and the bank charges 1.5% commission for each transaction.
6700 USD is exchanged for Swiss francs, and the money is placed in a savings account that pays a nominal rate of 6% p.a. compounded quarterly.

   i Calculate the number of Swiss francs to be invested.

   ii Find the amount in the account after 9 months.

e If the francs are exchanged back to US dollars after 9 months, how many dollars will there be?

f What assumption are you making in **e**?

10 **a** The first 3 terms of a sequence are 56, 28, and 14.

   i Show that the sequence is geometric.

   ii Find the 8th term of the sequence.

   iii Find the sum of the first 8 terms.

b The third term of another geometric sequence is 24.5 and the 5th term is 12.005. All of the terms of this sequence are positive.

   i Find the first term and the common ratio.

   ii Write down the general formula for a term of this sequence.

c The first n terms of the sequence in **b** are larger than the corresponding terms in the sequence $u_n = 20 \times (0.8)^{n-1}$. Find n.

d For the sequence $u_n = 20 \times (0.8)^{n-1}$, find the sum of the first:

   i 30 terms   **ii** 50 terms   **iii** 100 terms.
Give your answers to 3 significant figures in each case. Comment on your results.

11 **a** Misha takes out a loan to purchase a generator for his business. He borrows $20 000 at 12.5% per annum, compound interest. At the end of each year, Misha is required to pay k.

   i If Misha does repay k each year, explain why the amount owing on the loan at the end of the first year is $\$(20\,000 \times 1.125 - k)$.

   ii Write an expression, in terms of k, for the amount owing on the loan at the end of the second year.

   iii At the end of the second year, the amount owing on the loan will be $17 131.25. Find k.

b Misha paid a total of $24 000 for the generator. Its value depreciated after purchase so that, at the end of the first year, it was only worth $20 400. Its value continues to decline at the same rate.

   i Find the percentage decrease in the value of the generator in the first year.

   ii Calculate the value of the generator at the end of the second year.

   iii Write a formula for the value after n years.

   iv Sketch the graph of the value of the generator for the first 6 years.

12 **a** Consider the arithmetic sequence defined by $A_n = 19n - 13$. Write down:

   i the first term   **ii** the common difference.

b G_n is a geometric sequence with 3rd term 1920 and 10th term 15.

   i Calculate the common ratio for the sequence G_n.

   ii Show that $G_1 = 7680$.

c By considering the equation $A_n = G_n$, find the term which A_n and G_n have in common.

d Let S_{A_n} represent the sum of the first n terms of A_n, and S_{G_n} represent the sum of the first n terms of G_n.

   i Show that $S_{A_n} = \tfrac{19}{2}n^2 - \tfrac{7}{2}n$.

   ii Write down an expression for S_{G_n} in terms of n.

   iii Determine the minimum number of terms required for the sum of the first n terms of A_n to exceed the sum of the first n terms of G_n.

13 Eddie wants to have £5500 in 3 years' time to travel overseas. He deposits £1800 in an account paying 5.6% annual interest, compounding quarterly.

a   **i** Calculate the value of this investment after 3 years.

   ii Find the additional amount required for Eddie to reach his target.

b Eddie's father offers to add enough money to the original savings so that the final amount will be £5500. Find how much Eddie's father needs to add to the account.

c Eddie's friend Thomas is planning to travel overseas with him. Thomas is planning to borrow the whole £5500 and repay it by monthly instalments over the following 2 years. The bank will charge a rate of 9% p.a. simple interest.

  i Calculate the total amount that Thomas will need to repay.

  ii Determine the amount of each monthly payment.

d How much extra will Thomas pay for the trip compared to Eddie and his father?

14 Tahlia borrows $5000 from her bank. Her repayments each month are $250 plus the interest for that month. Interest is charged at 1.2% each month.

a Determine the payment Tahlia must make at the end of the:

  i first month   **ii** second month   **iii** third month.

b The monthly payments form an arithmetic sequence.
Find the first term and common difference for the sequence. Hence write a formula for the value of the nth payment for this loan.

c Calculate the size of the payment at the end of the tenth month.

d Determine the number of payments required for Tahlia to pay off the loan.

e Calculate the total amount Tahlia will pay for this loan.

15 A radioactive material loses 10% of its weight per year. The weight of the material at the start of the first year was 200 g.

a   **i** Write down the weight of the material at the start of the second and third years.

  ii The weights at the start of each year form a geometric sequence. Write down the common ratio of the sequence.

  iii Find the weight of the material remaining at the start of the 6th year.

  iv Sketch the graph of the weight of the material remaining for the first 6 years.

  v At the start of which year will the weight of the material fall below 20 g?

b The weight of a second sample of radioactive material decreased from 120 g to 49.152 g after 6 years. Find the annual percentage rate of decrease.

TOPIC 2:   DESCRIPTIVE STATISTICS

You should understand these terms:

- **Population**
  A defined collection of individuals or objects about which we want to draw conclusions.
- **Census**
  The collection of information from the **whole population**.
- **Sample**
  A subset of the population. It is important to choose a sample at **random** to avoid **bias** in the results.
- **Survey**
  The collection of information from a **sample**.
- **Data** (singular **datum**)
  Information about individuals in a population.

- **Parameter**
  A numerical quantity measuring some aspect of a population.
- **Statistic**
  A quantity calculated from data gathered from a sample. It is used to estimate a population parameter.
- **Categorical variable**
  A variable describing a particular quantity or characteristic. The data is divided into **categories**, and the information collected is called **categorical data**.
- **Quantitative variable**
  A variable which takes a numerical value. The information collected is called **numerical data**.
- A **quantitative discrete variable** takes exact number values and is often a result of **counting**.
- A **quantitative continuous variable** can take any numerical value within a certain range. It is usually a result of **measuring**.

DISCRETE DATA

- **Frequency table** - may include columns for tally and relative frequency.

Number	Tally	Frequency	Relative frequency
1	|	1	0.05
2	|||	3	0.15
3	₶||	8	0.4
4	₶	5	0.25
5	|||	3	0.15

- **Dot plot**

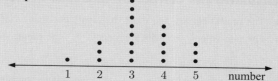

- **Column graph**

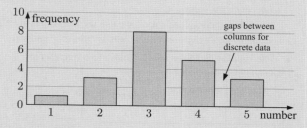

We can describe the distribution of a data set in terms of **skewness**.

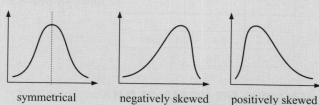

symmetrical   negatively skewed   positively skewed

Outliers are data values that are either much larger or much smaller than the general body of data.

If there are many different data values each with low frequencies, we **group** the data into **class intervals** with equal width.

CONTINUOUS DATA

For continuous data, the data is grouped into class intervals of equal width.

Instead of a column graph, we use a **frequency histogram** where there are no gaps between columns.

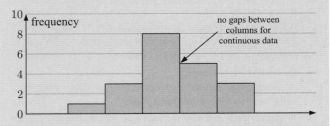

MEASURING THE CENTRE OF UNGROUPED DISCRETE DATA

- The **mode** is the most frequently occurring value.
- The **mean** is the arithmetic average of the data set.
  - ► The mean of a population is μ.
  - ► The mean of a sample is $\overline{x}$.

$$\overline{x} = \frac{\sum\limits_{i=1}^{n} x_i}{n}$$

- The **median** is the *middle value* of an ordered data set. If there are an even number of data, the median is the average of the two middle values.

  If there are n data values, the median is the $\left(\dfrac{n+1}{2}\right)$th value.

MEASURING THE CENTRE OF GROUPED DATA

- The **modal class** is the most frequently occurring class.
- For data values x_i with frequency f_i, the **mean** is

$$\overline{x} = \frac{\sum\limits_{i=1}^{k} f_i x_i}{n} \quad \text{where } n = \sum_{i=1}^{k} f_i \text{ and } k \text{ is the number of}$$

  different data values.
  For grouped continuous data, we use the mid-interval value of the class to represent all scores within that interval.
- The **median** is found using cumulative frequencies.

Number	Frequency	Cumulative frequency
1	4	4
2	5	9
3	10	19 ←
4	7	26
5	3	29
6	1	30
Total	30	

$\dfrac{n+1}{2} = 15.5$

15th and 16th values are both 3

MEASURING THE SPREAD OF DATA

The **range** is the difference between the maximum and minimum data values.

The **interquartile range** is the range of the middle half of the data.

$$\text{IQR} = Q_3 - Q_1$$

The **standard deviation** is found using technology.

- In this course we always use the formula for the *population standard deviation* σ.
- You must be able to calculate the standard deviation for both ungrouped and grouped data.

BOX AND WHISKER PLOTS

For box and whisker plots we need the five-number summary:

min, Q_1, median, Q_3, max.

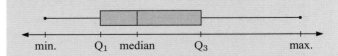

min. Q_1 median Q_3 max.

You should be able to use a **parallel boxplot** to compare two data sets.

CUMULATIVE FREQUENCY GRAPHS

These are constructed from a cumulative frequency table. We plot the cumulative frequencies, then join the points with a smooth curve.

We can use the graphs to calculate **percentiles**, which are scores below which a certain percentage of the data lies.

- The lower quartile Q_1 is the 25th percentile.
- The median Q_2 is the 50th percentile.
- The upper quartile Q_3 is the 75th percentile.

A percentile scale may be added to the cumulative frequency graph.

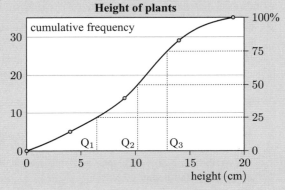

Height of plants

SKILL BUILDER - SHORT QUESTIONS

1 The numbers of customers entering a shop each hour on a particular day were: 14, 23, 26, 34, 24, 18, 26, 16, 25.

  a Is the data discrete or continuous?

  b Determine the mean, median, mode, and range for this data.

  c Find the total income for the shop if the mean amount spent per customer was $14.20.

2 The number of houses in each street of a council area is presented in the frequency table shown.

  a Is the data discrete or continuous?

  b Construct a column graph for this data.

  c What is the modal class for this data?

Houses	Frequency
0 - 9	5
10 - 19	12
20 - 29	14
30 - 39	18
40 - 49	8
Total	57

3 The numbers of birds counted in a park on a sample of 22 days last month were:

13, 22, 3, 14, 17, 7, 20, 18,
5, 22, 31, 25, 17, 35, 40, 36,
22, 27, 39, 34, 49, 42

a Determine the median and quartile values.

b Find the probability that, on any day, the number of birds present is more than the upper quartile.

4

Age (years)	1 - 5	6 - 10	11 - 15	16 - 20	21 - 25
Frequency	2	7	14	5	8

a Represent the distribution of ages as a column graph.

b Estimate the mean age.

c Write down the modal class.

5 The mean of the following set of numbers is 97.

90, 100, 93, 96, p, 107, 98, 98, 92

a Calculate the value of p.   **b** Find the median value.

6 Consider the grouped data alongside.

a Write down the modal class.

b Estimate the mean score.

c Draw a histogram to display this data.

Score (x)	Frequency (f)
$0 \leqslant x < 2$	2
$2 \leqslant x < 4$	7
$4 \leqslant x < 6$	15
$6 \leqslant x < 8$	11
$8 \leqslant x < 10$	5

7 The heights of 50 plants were measured, and the results presented in the table given.

a Is the data discrete or continuous?

b List the mid-interval values for each class.

c Estimate the mean height of the plants.

Height (cm)	Frequency
$0 \leqslant h < 10$	2
$10 \leqslant h < 20$	15
$20 \leqslant h < 30$	21
$30 \leqslant h < 40$	7
$40 \leqslant h < 50$	5
Total	50

d Write down the cumulative frequencies.

e Construct a cumulative frequency graph.

f Use your graph to determine the 80th percentile.

g How many plants are taller than the 80th percentile?

8 The column graph below shows the number of people living in each house on a suburban street.

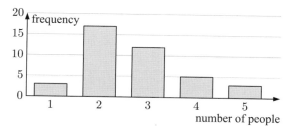

a Determine the number of houses on the street.

b Calculate the number of people who live on the street.

c Calculate the mean number of people per house.

d Find the percentage of houses which have more than 2 people.

9 The cumulative frequency curve below represents the finishing time, in minutes, of 30 competitors in a recent orienteering contest.

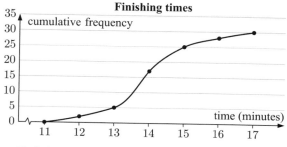

a Find the median finishing time, giving your answer to 1 decimal place.

b Find the time required for a runner to finish in the 1st quartile.

c How many runners finished in a time between 12 and 15 minutes?

10 **a** The mean of 7 integers is 14. In ascending order, the integers are 9, 10, a, 13, b, 16, 21.
Find the values of a and b.

b In ascending order, a set of six integers are: 1, 5, 9, 11, 16, p. The mean of the six numbers is the same value as their median. Find p.

11 The statistics below represent a sample of 30 employees' wages ($'000) at two companies.

Company 1:   mean = 38,   median = 35,
standard deviation = 7
Company 2:   mean = 38,   median = 41,
standard deviation = 11

a Which company has the greater dispersion in the wages paid to its employees?

b Which of the data sets is likely to have the smaller interquartile range?

c Which company has more highly paid employees?

12 A survey of the heights of Year 12 students at an international school gave the following results. All measurements have been rounded to the nearest 10 cm.

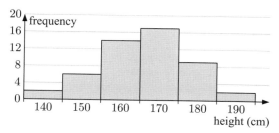

a What are the boundaries for the 170 cm class?

b Estimate the mean and standard deviation of the students' height in this survey.

c How many students were taller than 2 standard deviations above the mean?

13 The following data are the number of minutes taken for 12 contestants to complete a task on a TV game show.

20, 32, 15, 40, 26, 25, 19, 28, 21, 25, 16, 22

a Write down the mean and standard deviation for this information.

b A contestant whose time is more than one standard deviation above the mean is immediately eliminated. List the times of the people who are immediately eliminated.

c Recalculate the mean and standard deviation for the remaining times.

d The remaining contestants whose times lie within one standard deviation of the new mean will compete in the next game. Find the number of people who will contest the next game.

14 The frequency histogram illustrates the times taken by a group of people to solve a puzzle.

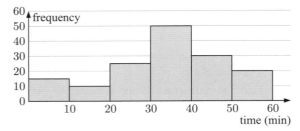

a Construct a cumulative frequency graph for the data.

b Hence estimate:
   i the median time taken to solve the puzzle
   ii the interquartile range of the data
   iii the probability that a randomly selected person was able to solve the puzzle within 35 minutes.

15 The times taken, in seconds, for 10 students to run 400 metres are:
$$55, 58, 61, 63, 67, 80, 83, 89, 94, 110$$

a Calculate:
   i the mean   **ii** the standard deviation.

b Greg adds his 400 metre running time to the results listed above. The new mean value is 75 seconds. What was Greg's 400 metre running time?

16 The list below shows the amount of weekly rent for a sample of studio apartments in southern Italy.

a Estimate the mean and standard deviation for the weekly rents.

b What is the probability that the rent for a randomly chosen studio apartment will be €140 or greater?

c Determine the percentage of studio apartments that have rents greater than one standard deviation above the mean.

Weekly rent (€)	Frequency
$80 \leqslant r < 100$	3
$100 \leqslant r < 120$	15
$120 \leqslant r < 140$	26
$140 \leqslant r < 160$	30
$160 \leqslant r < 180$	14
$180 \leqslant r < 200$	1

17 These parallel boxplots show the weights of particular species of fungi collected from 3 different sites in a forest.

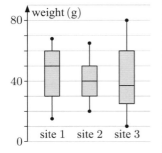

a Which site has the greatest range of weights?

b At which site are the weights of fungi least spread?

c Which site has the highest median weight of fungi?

d At which site were the heaviest fungi found?

e Which site has the highest proportion of weights above 40 grams?

18 Consider the ordered list of scores:
$$p, 15, 17, 19, q, q, r, r, r, 29, 33, 39$$

a If the range of the scores is 26, determine the value of p.

b The mean of the scores is 23 and the median is 22.
   i Write down two linear equations which could be used to determine the values of q and r.
   ii Solve your linear equations for q and r.

c State the modal score.

19 Consider the following boxplot:

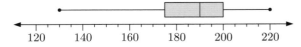

a Write down the values of the lower quartile, median, and upper quartile.

b Find the range of the data.

c Calculate the interquartile range.

20 The following marks were obtained by students in an examination:

67 62 75 78 78 49 57 59 61 72
75 25 82 68 85 81 48 70 76 87

a Find the five-number summary for this data.

b Represent this information in a box and whisker plot.

21 The numbers of spelling errors made by Year 9 students in an essay are displayed in this cumulative frequency curve.

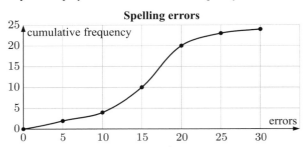

a Use the cumulative frequency curve to find the five-number summary for this data.

b Construct a box and whisker plot for the student spelling errors.

22 The prices (in £'000) of houses sold last month in a suburb were:
$$240, 260, 262, 280, 310, 325, 330, 340, 760$$

a Find the median price for these sales.

b Determine the interquartile range.

c If the score 760 is removed, what percentage change occurs in the median price?

23 The frequency table below shows the number of cars owned by different families.

Number of cars	Frequency (f)
0	78
1	117
2	69
3	18
4	2

a Is this discrete or continuous numerical data?

b Find the modal number of cars owned.

c How many families were in the group surveyed?

d For this data, calculate:

    i the mean   **ii** the standard deviation.

24 A cumulative frequency graph for the random variable X is given below.

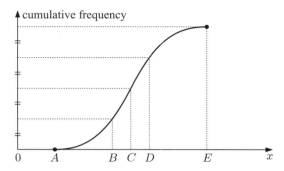

a What is represented by:

    i A **ii** B **iii** C **iv** D **v** E?

b What do these measure?

    i $E - A$ **ii** $D - B$

c Determine:

    i $P(B \leqslant X \leqslant D)$ **ii** $P(X \geqslant B)$

25 A class of 26 students measured the distance they travel to school each day. Their results are represented in the boxplot below.

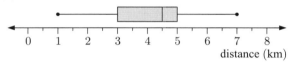

a Copy and complete:

Minimum	Q_1	Median	Q_3	Maximum

b State the interquartile range.

c Out of the 4 histograms below, which could represent the distance travelled to school data?

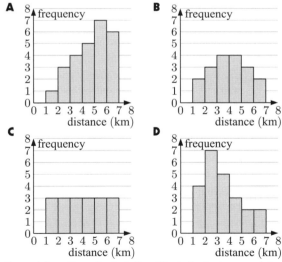

26 The times taken (in minutes) for 20 students to complete a problem are listed below:

    12.1   13.3   11.1   14.3   8.8   9.4   11.2

    14.9   11.7   11.1   10.4   11.2   13   12

    13.5   12.3   9.2   12.8   12.4   12.4

a Is this data best classified as 'discrete' or 'continuous' numerical data?

b Calculate the average time taken to solve the problem.

c What is standard deviation a measure of?

d Determine the standard deviation in times taken to complete the problem.

27 Consider the following scores, which are written in ascending order: 15, 17, p, p, p, 20, q, 26, 29, 30, 30, r

The mean score is 24, the median score is 22.5, and the mode is 19. Determine the values of p, q, and r.

28 The life of a battery is tested by measuring the length of time (in minutes) it can power a small motor.

a The life of a batch of *new batteries* is tested, and the results are as follows:

    19   21   23   24   24   25   26

    26   27   28   28   29   31

Using this data, complete the table below:

	Min.	Q_1	Median	Q_3	Max.
New batteries					

b Left unused, batteries lose their charge. The company that produces the type of battery tested above, claims "*our batteries retain more than* 85% *of their life after* 5 *years in storage*". To test this claim, a random sample of 15 unused batteries is tested after 5 years in storage. The summary statistics below outline the life of the *5 year old* batteries.

	Min.	Q_1	Median	Q_3	Max.
5 year old batteries	18	23	24	26	30

Based on the information provided, comment on the claim made by the manufacturer. Clearly state whether you agree or disagree with the claim, and give clear evidence to support your statement.

29 The frequency histogram below illustrates the lengths of time a student practices the piano each day over a period of 4 weeks.

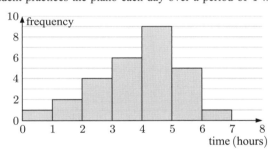

a What percentage of practices lasted between 3 and 4 hours?

b Draw a cumulative frequency graph to represent this information.

c Hence, estimate the median practice time.

30 The heights of 80 children from a kindergarten were measured. The results are shown in the table below.

Height (cm)	Number of students
$80 \leqslant h < 90$	8
$90 \leqslant h < 100$	12
$100 \leqslant h < 110$	17
$110 \leqslant h < 120$	30
$120 \leqslant h < 130$	13

a What is the mid-interval value for the class interval $80 \leqslant h < 90$?

b Estimate the:
  i mean height   **ii** standard deviation in height.
c What value is approximately 3 standard deviations above the mean?

31 A class of 24 students writes down the total number of children in their families:

    1 2 4 3 2 1 1 2 1 5 4 2
    2 1 3 2 3 4 1 2 1 2 2 1

a Is this data discrete or continuous?

b Construct a frequency table for the data.

c Determine the:
  i median value   **ii** mean value.

32 Before selecting a new mobile phone plan, George reviews the duration of calls he made over the last 3 months. George produced the histogram below to illustrate the data he collected.

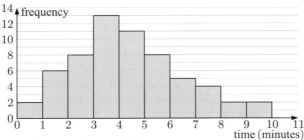

a Write down the modal class.

b Determine how many phone calls George made:
  i in total   **ii** per month.

c Estimate the probability that a call will last 6 minutes or more.

33 Jacob plants a number of small trees. At the time of planting, he measures the height of each planted tree.

The histogram below shows the distribution of tree heights.

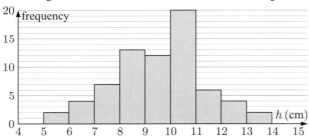

a Is the variable *height* discrete or continuous?

b How many trees did Jacob plant?

c Estimate the mean height, $\overline{h}$.

34 The cumulative frequency graph below represents the time taken for a number of competitors to run a race.

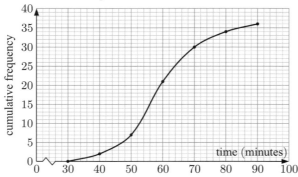

a How many competitors ran the race?

b Determine the median race time.

c The slowest race time was 85 minutes and the fastest was 31 minutes. Use this information to draw a boxplot for the race data.

35 The following table shows the frequency f of scores x.

x	4	5	6	7	8	9
f	5	7	3	2	1	1

a For the scores x, calculate the:
  i mean, $\overline{x}$ **ii** standard deviation, s.

b What percentage of the scores are:
  i 5 or less   **ii** greater than 5?

36 A hockey team records the number of goals scored for 15 matches. The mean number of goals (per match) was 2.4 .

a Is the variable *number of goals scored* discrete or continuous?

b Determine the total number of goals scored over the 15 matches.

c One more game is to be played to finish the season. How many goals must be scored in this final game to increase the mean number of goals to 3?

SKILL BUILDER - LONG QUESTIONS

1 A winemaker wants to examine the effect of weed spray in his vineyard. He randomly selects 50 sample spots, each of area 1 m^2, and counts the number of weeds in each spot. The results are shown in the table below.

Number of weeds	Frequency
0 - 4	9
5 - 9	15
10 - 14	10
15 - 19	9
20 - 24	5
25 - 29	2

a Construct a cumulative frequency table for the data.

b Using your table, determine the probability that a spot chosen at random from the 50 sample spots will have:
  i no more than 19 weeds   **ii** 9 or fewer weeds.

c Construct a cumulative frequency graph for the number of weeds. Use 2 cm to represent each class interval of weeds on the horizontal axis, and 1 cm to represent 10 sample spots on the vertical axis.

d Use your graph to estimate the:
  i median   **ii** first quartile   **iii** 75th percentile.

e Write down the interquartile range.

f Use your calculator to estimate the mean number of weeds per spot.

g The area of the vineyard is 12 000 m^2. Estimate the total number of weeds in the vineyard.

2 The prices in dollars, of 25 similar printers are:

    271 272 274 279 280 281 285 287 288
    290 290 292 293 294 297 298 299 299
    299 303 304 307 309 311 316

The mean of this data is $293 and the standard deviation is $12.

a Calculate the median, range, and lower and upper quartiles.

b Display these statistics on a horizontal box and whisker plot. Use a scale of 1 cm to represent $10.

c Three months later, the prices for the same printers were again recorded. They ranged between $269 and $329 with mean $295 and standard deviation $16. The lower and upper quartiles were $280 and $305 respectively, and the median was $295. Show this data as a box and whisker plot using the same scale as in **b**.

d Describe the main difference between the box and whisker plots.

e Explain whether this information shows that the price of printers has increased. Give a clear reason for your answer.

f Find the percentage increase in the mean price of printers over the 3 month period.

3 The distance travelled by two similar toy cars after rolling down a slope was measured 40 times each. The measurements were rounded to the nearest tenth of a metre.

Red car
3.6	4.6	5.6	6.4	4.2	5.3	6.1	4.5
5.4	4.6	3.9	6.2	5.8	4.5	5.4	6.1
4.5	5.6	5.7	4.8	3.9	5.6	6.1	5.9
4.1	5.3	4.2	6.2	7.4	5.4	5.8	4.5
3.9	5.4	5.7	4.8	5.4	5.7	6.1	6.4

Blue car
Number of rolls	40
Mean distance	4.9 m
Median distance	4.8 m
Shortest distance	3.2 m
Longest distance	6.7 m
First quartile	4.1 m
Third quartile	5.4 m
Standard deviation	0.8 m

a Determine the mean and standard deviation for the distance travelled by the red car.

b Complete the table of cumulative frequencies for the red car data.

Distance (m)	Cumulative frequency
$3.5 \leqslant d < 4.0$	
$4.0 \leqslant d < 4.5$	
$4.5 \leqslant d < 5.0$	
$5.0 \leqslant d < 5.5$	
$5.5 \leqslant d < 6.0$	
$6.0 \leqslant d < 6.5$	
$6.5 \leqslant d < 7.0$	
$7.0 \leqslant d < 7.5$	

c Draw the cumulative frequency graph for the distance travelled by the red car. Use a scale of 1 cm to represent 1 m on the horizontal axis and 1 cm to represent 10 units on the vertical scale.

d Use the graph to find the following statistics for the red car:
  i median distance   **ii** lower quartile
  iii upper quartile.

e Draw parallel box and whisker plots to display the data for both cars.

f Compare the statistics for distance travelled by the two toy cars. Is it reasonable to assume that the same machine manufactured these two toys? Give reasons for your answer.

4 A manufacturer states that the contents of its cereal boxes weigh an average of 320 g. A random sample of 24 boxes was weighed, with the following results recorded in grams:

312	320	326	330	306	322	326	330
312	308	307	316	315	328	334	309
308	325	320	332	316	321	314	324

a Calculate the mean weight and the range of weights for the boxes.

b Organise the data in a frequency table with the first class being 305 - 309.

c Use the information in your table to draw a frequency histogram for the cereal data.

d Comment on the manufacturer's stated average weight.

e Six months later, another randomly selected 24 boxes were weighed with the following results:

Weight (g)	Frequency
310 - 314	3
315 - 319	5
320 - 324	8
325 - 329	6
330 - 334	2

  i Calculate the mean weight for the new data.

  ii Does the evidence suggest that the manufacturer has improved the production process in the six months?

5 A large store employs 100 sales staff. The employees' total sales for last year are listed in the table below.

Sales ($'000)	Number of staff
60 to less than 70	3
70 to less than 80	8
80 to less than 90	11
90 to less than 100	23
100 to less than 110	27
110 to less than 120	20
120 to less than 130	8

a Represent this information as a frequency histogram.

b Write down the minimum and maximum sales required for a staff member to be in the highest class.

c Calculate the mean and the standard deviation of the sales per staff for the year using the midpoints of each class. Give your answers to the nearest hundred dollars.

d The store had an incentive scheme last year that offered a $900 bonus to all staff whose sales exceeded 2 standard deviations above the mean.

  i What was the minimum value of sales required for a staff member to qualify for this bonus?

  ii How much money did the store pay in bonuses?

e If the top 8 sales staff were to get the bonus, how many standard deviations above the mean would the limit need to be set to?

f At the end of the year the store manager decides to reduce the number of sales staff. Every staff member whose sales were 1.385 standard deviations below the mean or lower will be removed from the team. How many staff will lose their jobs?

6 The table below shows the length (in seconds) of a number of songs on a portable music player.

Duration (s)	Frequency	Cumulative frequency
$0 \leqslant t < 60$	5	5
$60 \leqslant t < 120$	20	25
$120 \leqslant t < 180$	40	x
$180 \leqslant t < 240$	60	y
$240 \leqslant t < 300$	25	150
$300 \leqslant t < 360$	10	160

a Write down the modal class.

b Determine the values of x and y.

c On graph paper, draw a cumulative frequency curve to illustrate this data. Use 2 cm to represent 60 seconds on the horizontal axis, and 1 cm to represent 10 songs on the vertical axis.

d Using your cumulative frequency curve, estimate:
  i the median song length   **ii** the interquartile range.

e If a song is selected at random, estimate the probability that the song is no longer than 4.5 minutes.

f Given that the shortest song length was 40 seconds and the longest 340 seconds, draw a box and whisker plot for the song length data. Use a scale of 2 cm represents 60 seconds.

7 Employees in a law firm often work overtime. In a given week, the number of hours overtime worked by the employees were:

0, 0, 1, 2, 4, 4, 4, 5, 5, 5, 5, 6, 6,
7, 9, 9, 11, 11, 12, 13, 15, 18, 22, 26, 34

a Calculate the value of:
  i Q_1 **ii** the median   **iii** Q_3

b Draw a boxplot to illustrate the data.

c Based on your boxplot, comment on whether the data appears symmetric or not. Explain your answer.

d For the given data, write down the:
  i range   **ii** interquartile range.

e Complete the following statements about the data:
  i *"The central 50% of employees work between ...... and ...... hours of overtime."*
  ii *"75% of employees work a maximum of approximately ...... hours overtime."*

f The median, the mode, and the mean each measure the 'centre' of a data set. Which measure of 'centre' would the law firm choose to suggest its employees do not work as much overtime?

8 The lengths (in minutes) of a number of games in a chess tournament are recorded. The cumulative frequency graph below illustrates the game length information.

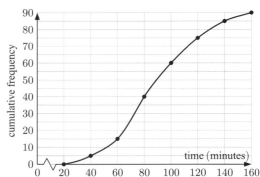

a How many games were played during the chess tournament?

b Find the median game length.

c Estimate the interquartile range for the data.

d Estimate the 10th percentile for the data.

e Draw a frequency histogram to represent the game length data.

f Hence, estimate the:
  i mean value   **ii** standard deviation.

9 The volume v, of milk consumed by an infant over a large number of feeds is shown in the table below.

Volume, v (in mL)	Frequency	Cumulative frequency
$80 \leqslant v < 100$	5	5
$100 \leqslant v < 120$	15	20
$120 \leqslant v < 140$	25	45
$140 \leqslant v < 160$	40	85
$160 \leqslant v < 180$	60	145
$180 \leqslant v < 200$	20	165
$200 \leqslant v < 220$	5	170

a How many feeds are shown in the table?

b Draw a cumulative frequency graph to represent this data.

c Hence, estimate the:
  i median   **ii** interquartile range.

d Use technology to estimate the:
  i mean volume, $\overline{v}$ **ii** standard deviation, s.

e Let $p = \overline{v} - s$ and $q = \overline{v} + s$.
  i Write down the values of p and q.
  ii Using your cumulative frequency graph, estimate the probability that the infant drinks between p and q mL in a single feed.

TOPIC 3:   SETS, LOGIC, AND PROBABILITY

SETS

A **set** is a collection of numbers or objects.

The numbers or objects in a set are called **elements** or **members** of the set.

The **empty set** { } or $\varnothing$ contains no elements.

The number of elements in set A is $n(A)$.

A set which has a finite number of elements is called a **finite set**.

An **infinite set** has infinitely many elements.

The **universal set** U is the set of all elements under consideration.

P is a **subset** of Q if every element of P is also an element of Q.	$P \subseteq Q$
P is a **proper subset** of Q if P is a subset of Q but is not equal to Q.	$P \subset Q$
The **intersection** of P and Q consists of all elements which are in both P and Q.	$P \cap Q$
The **union** of P and Q consists of all elements which are in P or Q or both.	$P \cup Q$
P and Q are **disjoint** or **mutually exclusive** if they have no elements in common.	$P \cap Q = \varnothing$
The complement of P, denoted P', is the set of all elements of U which are *not* in P.	$P \cap P' = \varnothing$ $P \cup P' = U$

VENN DIAGRAMS

A **Venn diagram** consists of a universal set U represented by a rectangle.

Sets within the universal set are usually represented by circles.

Subsets, intersection, union, and complement can be represented by shaded regions on a Venn diagram.

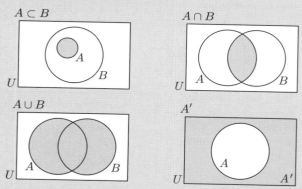

You should be able to interpret the meaning of different shaded regions of a Venn diagram.

You should be able to use Venn diagrams to:

- verify set identities
- solve problems regarding the number of elements in sets.

LOGIC

Propositions are statements which may be true or false. The **truth value** of a proposition is whether it is true or false.

Propositions may be **indeterminate**, which means their truth value may be different depending on the individual they are applied to.

Compound propositions are statements which are formed using the connectives *and* and *or*.

You should understand the following terms relating to propositions p and q, which have corresponding regions P and Q on a Venn diagram.

Expression	Meaning	Notation	Region on Venn diagram
negation of p	not p	$\neg p$	P'
conjunction of p and q	p and q	$p \wedge q$	$P \cap Q$
disjunction of p and q	p or q	$p \vee q$	$P \cup Q$
exclusive disjunction of p and q	p or q but not both	$p \veebar q$	$(P \cap Q') \cup (P' \cap Q)$

You should be able to draw regions to display each of these terms on a Venn diagram.

Truth Tables

p	q	$\neg p$	$\neg q$	$p \wedge q$	$p \vee q$	$p \veebar q$
T	T	F	F	T	T	F
T	F	F	T	F	T	T
F	T	T	F	F	T	T
F	F	T	T	F	F	F

You should understand these terms relating to compound propositions:

- **Logical equivalence:** propositions having the same truth table column.
- **Tautology:** compound propositions whose truth table column is all true.
- **Logical contradiction:** compound propositions whose truth table column is all false.

- **Validity:** an argument is valid if and only if it is a tautology.

You should be able to construct truth tables involving three propositions.

Further Terminology

- The **implication** $p \Rightarrow q$ means "if p then q". p is called the **antecedent** and q is called the **consequent**.
- The **equivalence** $p \Leftrightarrow q$ means "p if and only if q".
- The **contrapositive** of $p \Rightarrow q$ is $\neg q \Rightarrow \neg p$.
- The **converse** of $p \Rightarrow q$ is $q \Rightarrow p$.
- The **inverse of** $p \Rightarrow q$ is $\neg p \Rightarrow \neg q$.

p	q	$p \Rightarrow q$	$p \Leftrightarrow q$	$q \Rightarrow p$	$\neg p \Rightarrow \neg q$	$\neg q \Rightarrow \neg p$
T	T	T	T	T	T	T
T	F	F	F	T	T	F
F	T	T	F	F	F	T
F	F	T	T	T	T	T

PROBABILITY

The chance or likelihood of an event happening is described by a **probability**.

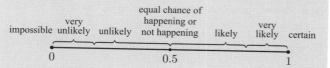

Probability Experiments

The **number of trials** is the number of times the experiment is repeated.

The **outcomes** are the different results possible for each trial.

The **frequency** of an outcome is the number of times this outcome is observed.

The **relative frequency** of an outcome is its frequency expressed as a fraction or percentage of the number of trials.

The **experimental probability** of an event is the relative frequency with which the event is observed.

A **sample space** U is the set of all possible outcomes of an experiment.

You should be able to display a sample space by:

- listing outcomes
- tables of outcomes
- Venn diagrams.
- 2-dimensional grids
- tree diagrams

Theoretical Probability

For an experiment with equally likely outcomes, event A has probability

$$P(A) = \frac{n(A)}{n(U)}$$

Two events are **complementary** if exactly one of the events *must* occur.

For an event A, the complementary event is A', and

$$P(A') = 1 - P(A).$$

Laws of Probability

- $P(A \cup B) = P(A) + P(B) - P(A \cap B)$.
- **Mutually exclusive events** are events which share no common outcomes. In this case $P(A \cap B) = 0$.

- **Conditional probability**

  The probability of A occurring given that B has occurred is

  $$P(A \mid B) = \frac{P(A \cap B)}{P(B)}$$

- Two events are **independent** if the occurrence of one of the events does not affect the probability that the other occurs.
  If A and B are independent then
  $$P(A \cap B) = P(A) \times P(B).$$
  If A and B are dependent then
  $$P(A \text{ then } B) = P(A) \times P(B \mid A).$$

Using Diagrams to Find Probabilities

You should be able to find probabilities using tree diagrams, Venn diagrams, and 2-dimensional grids.

When using tree diagrams:

- the probability of each outcome is obtained by *multiplying* the probabilities along its branch
- if there is more than one outcome in an event, we *add* the probabilities of the outcomes.

Sampling

You should understand how sampling is affected by the **replacement** of objects between samples.

Sampling **with replacement** leads to independent events.

Sampling **without replacement** leads to dependent events.

SKILL BUILDER - SHORT QUESTIONS

1 Suppose $U = \{$natural numbers less than 12$\}$,
$A = \{$multiples of 3$\}$, and $B = \{$factors of 10$\}$.
List the elements of: **a** $A \cap B$ **b** $A \cup B$ **c** $(A \cup B)'$.

2 Set $A = \{$multiples of 4 greater than 0 and less than 20$\}$
Set $B = \{$even numbers greater than 0 and less than 10$\}$

  a Find $A \cup B$. **b** Find $A \cap B$.
  c Set C contains 3 elements and is a subset of A. Write down one possible set C.

3 $\mathbb{N}$ is the set of natural numbers, $\mathbb{Z}$ is the set of integers, and $\mathbb{Q}$ is the set of rational numbers.

  a Write down an element that is in:
   i $\mathbb{Z}$ **ii** $\mathbb{Z}'$ **iii** $\mathbb{N} \cap \mathbb{Q}$ **iv** $\mathbb{Q}' \cap \mathbb{Z}$
  b Draw a Venn diagram to represent the sets $\mathbb{N}$, $\mathbb{Z}$, and $\mathbb{Q}$.

4 Let $U = \{$positive integers greater than 7 and less than 19$\}$,
$A = \{$multiples of 3$\}$, and $B = \{$factors of 36$\}$.

  a List the elements of $A \cap B$.
  b List the elements of A'.
  c Represent the relationship between sets A and B on a Venn diagram.
  d Are the following statements true or false?
   i $B \subset A$ **ii** $n(A' \cap B') = 7$ **iii** $A \cup B = A$

5 Let $U = \{$positive integers less than 20$\}$,
$P = \{$prime numbers less than 20$\}$, and
$F = \{$factors of 24$\}$.
  Find: **a** $n(F)$ **b** $P \cap F$ **c** $P \cup F$ **d** $P' \cap F$

6 On separate Venn diagrams like the one provided, shade the region specified:

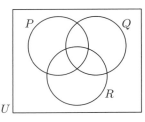

  a P' **b** $P \cup R$
  c $P' \cap Q$
  d $P \cap (Q \cup R)$

7 The Venn diagram shows the number of students in a year group who like certain foods.

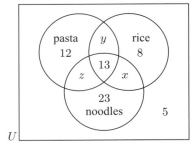

  a Write down the number of students who like:
   i pasta **ii** rice and noodles but not pasta
   iii at least two of the foods.
  b Determine the probability that a student likes:
   i none of these foods
   ii rice, given that the student likes noodles.

8 Suppose $U = \{$positive integers $< 10\}$,
$P = \{$even numbers $< 10\}$, and
$Q = \{$prime numbers $< 10\}$.

  a Write down the members of:
   i $P \cap Q$ **ii** $(P \cup Q)'$
  b Represent the given information on a Venn diagram.

9 Consider the propositions: p: Toby has four legs.
   q: Toby has long ears.
   r: Toby is a dog.

  a Write in symbols: "If Toby is not a dog and Toby has long ears, then Toby does not have four legs."
  b Write in words the converse of the statement $q \vee \neg p \Rightarrow \neg r$.

10 Suppose $U = \{n \mid 1 \leqslant n \leqslant 12, \ n \in \mathbb{N}\}$,
$P = \{$odd numbers$\}$, $Q = \{$factors of 12$\}$, and
$R = \{$multiples of 5$\}$.
List the elements of:
  a $P \cap Q$ **b** $(P \cap Q) \cup R$ **c** $Q \cap R$ **d** $P' \cap (Q \cup R)$

11 The Venn diagram below shows the number of different types of movies available for hire.
$P = \{$comedy$\}$, $Q = \{$romance$\}$, $R = \{$adventure$\}$

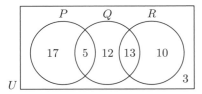

Find:
  a $n(P)$ **b** $n(P \cup Q)$ **c** $n(Q \cap R)$
  d $n((P \cup Q \cup R)')$ **e** $n(Q')$ **f** $n(P \cap R)$

12 **a** A number of IB students study English or Spanish, or both English and Spanish. 25 study Spanish and 18 study English. If 6 students study both languages, how many IB students are there?

b An international school offers its programme in both French and English languages. 60% of the students take lessons in the English language programme and 76% take lessons in the French language programme. What percentage of the students take lessons in both languages?

c In a class of 25 students, 20% do not study art or drama. 13 students study art, and 9 students study drama. How many students study both art and drama?

13 Consider the following propositions:
p: Peter has black hair. q: Peter plays baseball.
r: Peter is a student.

a Write each of the following arguments in symbols.

i If Peter does not play baseball then he is not a student.

ii If Peter does not have black hair then he is neither a student nor does he play baseball.

b Write the following argument in words:
$$\neg r \Rightarrow \neg(p \vee q).$$

14 The Venn diagram below shows the number of students in a class who study various subjects.
$M = \{$students who study Music$\}$,
$P = \{$students who study Physics$\}$,
$D = \{$students who study Drama$\}$.

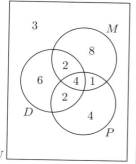

Find the number of students who:

a study Physics   **b** study Music or Drama but not both

c study Physics but not Music

d do not study Drama.

15 Let $U = \{x \mid 1 < x < 50, \; x \in \mathbb{Z}\}$.
A, B, and C are subsets of U such that $A = \{$factors of 48$\}$,
$B = \{$multiples of 6$\}$, and $C = \{$multiples of 8$\}$.

a List the elements in $A \cap B \cap C$.

b Find $n(B)$.   **c** List the elements in $A' \cap B$.

16 The sets X, Y, and Z are all subsets of the universal set $U = \{0, 1, 2, 3, 4, 5, 6, 7, 8, 9\}$.

The sets are defined as: $X = \{x \mid 0 \leqslant x \leqslant 4, \; x \in \mathbb{Z}\}$
$Y = \{y \mid y = 3 + x, \; x \in X\}$
$Z = \{z \mid z = 2x, \; x \in X\}$

a Write down the elements of set:

i Y **ii** Z

b Draw a Venn diagram showing the relationship between the sets X, Y, and Z. Include all members of set U in their appropriate positions.

17 Consider the propositions:
p: Wilson is an active dog.
q: Wilson digs holes.

a Write down the following propositions using words only:

i $\neg p \wedge \neg q$ **ii** $p \veebar q$ **iii** $\neg p \Rightarrow q$

b By completing this truth table, state whether the argument in **a iii** is a tautology, a logical contradiction, or neither.

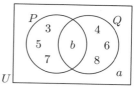

p	q	$\neg p$	$\neg p \Rightarrow q$
T	T	F	
T	F	F	
F	T	T	
F	F	T	

18 In the Venn diagram opposite,
$U = \{x \mid 1 < x < 10, \; x \in \mathbb{N}\}$
and $P = \{$prime numbers $\in U\}$

a Find the element b.

b Describe in words the set Q.

c Find the element a.

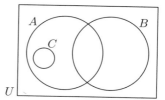

19 Consider the statement $p \Rightarrow q$ for:
p: I like the beach.
q: I live near the sea.
For this statement, write in symbols and words:

a the converse   **b** the inverse

c the contrapositive.

20 The Venn diagram shows the sets U, A, B, and C. Each region contains at least one element.

a State whether the following propositions are true or false:

i $A \cap B = \varnothing$

iii $B \subset A'$

ii $A \cap C = C$

iv $C \subset (A \cap B)$

b Shade the region $A' \cap B$ on the Venn diagram.

21 The Venn diagram below shows the number of members at a sports club who take part in various sporting activities.
$S = \{$members who play soccer$\}$,
$B = \{$members who play basketball$\}$,
$G = \{$members who play golf$\}$.

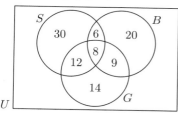

Find the number of members who:

a play all 3 sports

b play soccer or basketball but not both

c play only golf

d do not play basketball

e play more than one sport

f play soccer but do not play golf.

22 Let p and q be the propositions:
p: The sun is shining.
q: I take the dog for a walk.

a Write the following propositions using words only:

i $p \Rightarrow q$ **ii** $\neg p \vee q$

b Write, in words, the converse of $p \Rightarrow q$.

c Write in symbolic form:
"If I do not take the dog for a walk then the sun is not shining."

d Is the proposition in **c** the inverse, converse, or contrapositive of $p \Rightarrow q$?

23 Let p and q be the propositions:

  p: Molly has a DVD player.
  q: Molly likes watching movies.

  a Write the following proposition using words only:
  $\neg p \Rightarrow \neg q$.

  b Complete the following truth table for $q \Rightarrow p$.

p	q	$q \Rightarrow p$
T	T	
T	F	
F	T	
F	F	

  c Complete the following truth table for $\neg p \Rightarrow \neg q$.

p	q	$\neg p$	$\neg q$	$\neg p \Rightarrow \neg q$
T	T	F	F	
T	F	F	T	
F	T	T	F	
F	F	T	T	

  d Are the two propositions $q \Rightarrow p$ and $\neg p \Rightarrow \neg q$ logically equivalent? Give a reason for your answer.

24 Consider the two propositions p and q.

  a Complete the truth table for the compound proposition $(p \wedge q) \Rightarrow p$.

p	q	$p \wedge q$	$(p \wedge q) \Rightarrow p$
T	T	T	
T	F		
F	T		
F	F	F	

  b Is the compound proposition $(p \wedge q) \Rightarrow p$ a logical contradiction, a tautology, or neither?

25 Out of every 500 packets produced by a machine, 24 packets contain more than 40 biscuits. If two packets are chosen at random with replacement, find the probability that:

  a both contain more than 40 biscuits
  b one contains more than 40 biscuits and the other contains forty or less.

26 Consider the two propositions p and q.
Complete the truth table below for the compound proposition $(p \Rightarrow \neg q) \vee (\neg p \Rightarrow q)$.

p	q	$\neg q$	$p \Rightarrow \neg q$	$\neg p$	$\neg p \Rightarrow q$	$(p \Rightarrow \neg q) \vee (\neg p \Rightarrow q)$
T	T	F			F	
T	F	T			F	
F	T	F			T	
F	F	T			T	

State whether the result above is a logical contradiction, a tautology, or neither.

27 Three propositions are defined as follows:

  p: The weather is hot. q: I have money.
  r: I buy an ice cream.

  a Describe, in words only, the following propositions:
   i $(\neg r \wedge \neg q) \vee p$ **ii** $(p \wedge q) \Rightarrow r$
   iii $(\neg p \vee \neg q) \Rightarrow \neg r$

  b Write, in words, the contrapositive for the proposition $p \Rightarrow r$.

28 Consider the universal set $U = \{x \mid 0 \leqslant x \leqslant 9, \ x \in \mathbb{Z}\}$ containing the two subsets $P = \{\text{prime numbers}\}$ and $Q = \{\text{multiples of 3}\}$.

  a Draw a Venn diagram showing the relationship between P and Q. Clearly indicate the position of all elements in U.

  b Determine the probability that a randomly selected element of U is:
   i a multiple of 3
   ii a prime number and not a multiple of 3
   iii a multiple of 3 given that it is prime.

29 Consider the two propositions:

  p: I eat an apple every day. q: I visit the doctor.

  a State the negation of q.

  b In words only, write the following propositions:
   i $\neg p \Rightarrow q$ **ii** $(\neg p \vee q) \Rightarrow q$.

  c For the compound proposition $(\neg p \vee q) \Rightarrow q$, state whether it is a tautology, a logical contradiction, or neither.

p	q	$\neg p$	$\neg p \vee q$	$(\neg p \vee q) \Rightarrow q$
T	T	F	T	T
T	F	F		T
F	T	T	T	
F	F	T		

30 **a** Complete the truth table below for the compound proposition $(a \Leftrightarrow b) \wedge b \Rightarrow a$.

a	b	$a \Leftrightarrow b$	$(a \Leftrightarrow b) \wedge b$	$(a \Leftrightarrow b) \wedge b \Rightarrow a$
T	T			
T	F			
F	T			
F	F			

  b What word best describes the compound proposition $(a \Leftrightarrow b) \wedge b \Rightarrow a$?

31 **a** Complete the truth table below for the compound proposition $\neg(p \wedge q) \wedge q \Rightarrow (\neg p \vee q)$.

p	q	$p \wedge q$	$\neg(p \wedge q)$	$\neg(p \wedge q) \wedge q$	$\neg p \vee q$	$\neg(p \wedge q) \wedge q \Rightarrow (\neg p \vee q)$
T	T					
T	F					
F	T					
F	F					

  b What word best describes the compound proposition $\neg(p \wedge q) \wedge q \Rightarrow (\neg p \vee q)$?

32 The probability of rain on any day in December is 0.2.
The partially completed tree diagram below shows the possible outcomes when the weather for two consecutive days is considered.

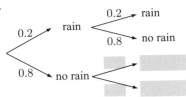

  a Fill in the boxes to finish the tree diagram.

  b What assumption are you making in **a**?

  c Use the tree diagram to determine the probability of it raining:
   i two days in a row   **ii** on one day only.

33 In the Venn diagram below, sets A, B, and C are subsets of the universal set $U = \{\text{positive integers less than } 10\}$.

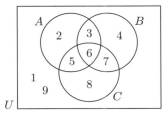

a Find:
  i $n(A)$ **ii** $n(A \cap B)$

b List the elements in:
  i $A \cup B \cup C$ **ii** $(A' \cap B) \cup C$

34 Propositions p and q are defined as:
  p: The sink is blocked.
  q: The plumber will clear the drain.

a Write in words:
  i $p \Rightarrow q$ **ii** the contrapositive of $p \Rightarrow q$.

b Use truth tables to determine whether the statements in **a** are logically equivalent.

35 A die with 4 red faces and 2 green faces is rolled twice.

a Copy and complete the following tree diagram to illustrate the possible outcomes.

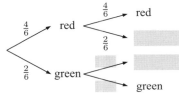

b Use the tree diagram to determine the probability of rolling:
  i two reds   **ii** two greens   **iii** one of each colour.

36 Let $U = \{\text{all letters in the English alphabet}\}$,
  $X = \{\text{all letters in the word "mathematics"}\}$, and
  $Y = \{\text{all letters in the word "intelligence"}\}$.

a Write down all unique elements of:
  i X **ii** Y

b Draw a Venn diagram which illustrates the relationship between X and Y. Show the elements in U which belong to sets X and Y.

c A letter is randomly chosen from the alphabet. Determine the probability that the letter selected is:
  i part of either of the words "mathematics" or "intelligence"
  ii part of the word "mathematics" but not part of the word "intelligence".

37 Events A and B have the following probabilities:
  $P(A) = 0.35$, $P(B) = 0.7$, $P(A \cup B) = 0.8$.

a Calculate $P(A \cap B)$.

b Represent this information on a Venn diagram.

c Find $P(A' \cap B')$.

d State, with a reason, whether events A and B are independent.

38 a Copy and complete the truth table:

p	q	$\neg p$	$\neg q$	$p \Rightarrow q$	$(p \Rightarrow q) \wedge \neg q$	$(p \Rightarrow q) \wedge \neg q \Rightarrow \neg p$
T	T					
T	F					
F	T					.
F	F					

b Consider the two propositions:
  p: Adrian works hard. q: Adrian earns money.
  Comment on the validity of the statement:
  "*If Adrian works hard then he earns money. Adrian does not earn money, so he is not working hard.*"

39 The Venn diagram alongside shows the number of students in a class that have dogs (D) or cats (C) for pets. Determine the number of students:

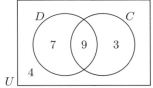

a in the class

b who have both cats and dogs

c who only have dogs

d who have at least one of the pets

e who have no pets.

40 If Katie wakes up late, the probability she will get to work on time is 0.2.
If Katie wakes up on time, the probability that she will be late for work is 0.2.

a Draw a tree diagram which represents this information.

b Suppose Katie wakes up late 30% of the time. Calculate the probability that she will be:
  i late for work tomorrow
  ii on time for work tomorrow.

41 The Venn diagram shows the probabilities between two events A and B.

a Find the value of the probability c.

b Use the Venn diagram to determine:
  i $P(A \cup B)$ **ii** $P(A \cap B)$ **iii** $P(A' \cap B)$
  iv $P(A \mid B)$ **v** $P(B \mid A)$

42 Let a and b be the propositions:
  a: The lights are on. b: Somebody is at home.

a Write the following compound propositions in words:
  i $a \Rightarrow b$ **ii** $\neg b \wedge \neg a$ **iii** $a \veebar b$

b Write down the statement which corresponds to:
  i the inverse of $a \Rightarrow b$
  ii the contrapositive of $a \Rightarrow b$.

c Using a truth table, show that the inverse and converse of $a \Rightarrow b$ are logically equivalent.

43 The table alongside shows the different types of chocolates in a packet.

	soft	medium	hard
dark	10	24	15
light	18	14	9

a One chocolate is chosen at random. Find the probability that the chocolate is:
  i dark or hard   **ii** light, given that it is soft.

b Two chocolates are chosen at random, without replacement. Find the probability that:

   i both are medium   **ii** at least one is soft and light.

44 **a** Copy and complete the truth table below:

p	q	$p \wedge q$	$p \veebar q$	$(p \wedge q) \vee (p \veebar q)$
T	T			
T	F			
F	T			
F	F			

b Given that $x \in \mathbb{R}$, consider the two propositions:

$$p:\ x \geqslant 1 \qquad q:\ x^2 - 16 = 0$$

For what value(s) of x are the following propositions true:

  i q **ii** $p \wedge q$ **iii** $(p \wedge q) \vee (p \veebar q)$?

45 Consider the table alongside. Find:

	A	B	C
W	7	3	11
Z	4	5	2

  a $n(A \cap Z)$ **b** $n(A \cup B)$

  c $n((B \cap W) \cup (C \cap Z))$

  d $P(A \cap B)$ **e** $P((C \cap W) \,|\, (B \cup C))$

46 Consider the following argument:

If Kania cooks the meal then it will be delicious.

If the meal is delicious then Mal will be happy.

Mal is happy.

Therefore, Kania did cook the meal.

  a Identify the three propositions (p, q, r) which form the argument.

  b Write the argument in symbolic form.

  c Use a truth table to determine the validity of the argument in **b**.

SKILL BUILDER - LONG QUESTIONS

1 The sets P, Q, and R are all subsets of U.

$U = \{x \,|\, 1 \leqslant x \leqslant 15,\ x \in \mathbb{Z}\}$, $P = \{\text{prime numbers}\}$,

$Q = \{\text{multiples of } 5\}$, $R = \{\text{factors of } 15\}$.

  a Write down the elements of:

    i P **ii** Q **iii** R

  b Hence, write down the elements of:

    i $P \cap Q \cap R$ **ii** $(P \cup Q \cup R)'$

  c Copy and complete the Venn diagram, placing all of the elements of U in the appropriate position.

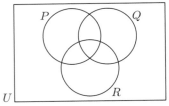

  d An element of U is randomly selected. Determine the probability that the value selected is a member of:

    i Q **ii** R' **iii** $Q \cup R$ **iv** $P \cap (Q \cup R)'$

  e In a sentence, give a reason why $n(P \cap Q \cap R') = 0$.

2 Let: $U = \{x \,|\, x \leqslant 10,\ x \in \mathbb{N}\}$, $P = \{\text{multiples of } 2\}$,

$Q = \{\text{multiples of } 3\}$, $R = \{\text{factors of } 12\}$.

  a List the elements of:

    i P **ii** Q **iii** R **iv** $P \cap Q \cap R$

  b Draw a Venn diagram to show the relationship between sets P, Q, and R. Show all of the elements of U.

  c Describe in words the sets:

    i $P \cup Q$ **ii** $P' \cap Q' \cap R'$

d Let p, q, and r be the statements:

  p: x is a multiple of 2. q: x is a multiple of 3.

  r: x is a factor of 12.

  i On a Venn diagram, shade the region corresponding to $p \veebar q$.

  ii Use a truth table to find the values of $(p \wedge r) \Rightarrow (p \veebar q)$.

    Write the first three columns of your truth table as shown opposite:

p	q	r
T	T	T
T	T	F
T	F	T
T	F	F
F	T	T
F	T	F
F	F	T
F	F	F

  iii Write down an element of U for which $(p \wedge r) \Rightarrow (p \veebar q)$ is true.

3 Consider two overlapping sets P and Q.

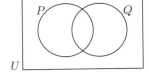

  a On separate Venn diagrams, shade the regions corresponding to:

    i $p \veebar q$ **ii** $\neg(p \wedge q)$

    iii $\neg p \vee \neg q$

  b By referring to **a ii** and **a iii**, explain whether $\neg(p \wedge q)$ and $\neg p \vee \neg q$ are logically equivalent.

  c Consider p: The bush has thorns.

            q: The bush is a rose bush.

  Write a sentence for the proposition $\neg(p \wedge q) \Rightarrow \neg q$.

  d Construct a truth table for the compound proposition in **c**. Begin by writing the first two columns of your truth table as shown opposite. Comment on your result.

p	q
T	T
T	F
F	T
F	F

4 A group of 20 food critics attended a restaurant and were served an entreé, a main course, and a dessert. Each critic recorded which of the three courses they enjoyed. Of the 20 critics:

  5 critics enjoyed all three courses,

  3 critics enjoyed only the entreé,

  1 critic enjoyed only the main,

  2 critics enjoyed only the dessert,

  6 enjoyed the entreé and the main,

  9 enjoyed the main and the dessert,

  7 enjoyed the entreé and the dessert.

  a Let E represent the set of critics who enjoyed the entreé, M represent the set of critics who enjoyed the main course, and D represent the set of critics who enjoyed the dessert. Represent the information on a Venn diagram, clearly labelling each region.

  b Using your Venn diagram, determine the number of critics who enjoyed:

    i exactly one course   **ii** two or more courses

    iii none of the courses.

  c If a critic is selected at random, determine the probability that the critic is a member of:

    i M **ii** $M \cap E$ **iii** $E \cup D$

  d The names of each of the 20 critics are placed in a hat. Two names are randomly selected without replacement. Determine the probability that:

    i both critics selected enjoyed the dessert

    ii at least one of the critics selected enjoyed the dessert

    iii both critics selected enjoyed the dessert given that at least one did.

5 Consider the positive integers less than 10.

Let p: x is an odd number. q: x is a multiple of 3.
   r: x is a prime number greater than 2.

a Write in symbolic language:

i If x is a prime number greater than 2 then x is an odd number.

ii If x is a multiple of 3 and x is an odd number, then x is a prime number greater than 2.

b Write down the truth sets for the following compound statements:

i $p \wedge r$ **ii** $p \veebar r$ **iii** $(\neg p \wedge q) \vee r$

c Consider the statement $(p \wedge \neg q) \Rightarrow \neg r$.

i Describe in words what the statement means.

ii Construct a truth table for the statement.

iii List the truth values of p, q, and r which make the statement false.

iv Write down a value of x which illustrates your answer in **iii**.

6 92 students at a school are enrolled in the IB Diploma programme.

20 of the students study Biology and Economics.
14 study Biology and Art.
10 study Economics and Art.
12 study Art but neither Biology nor Economics.
15 study Economics but neither Art nor Biology.
27 study none of these subjects.

a If there are 38 students who study Economics, determine the number of students who study all 3 subjects.

b Write down the number of students who study Biology but neither Art nor Economics.

c Draw a Venn diagram to represent all the information above.

d Find the probability that a randomly chosen student studies:

i Economics

ii Art, given that they study at least one of these subjects.

e Two students are chosen at random. Determine the probability that both students study:

i Biology

ii exactly two of these subjects.

7 Consider two propositions p and q and the compound proposition $\neg(p \vee q) \Rightarrow \neg p \wedge \neg q$.

a **i** Copy and complete the truth table below.

p	q	$\neg p$	$\neg q$	$p \vee q$	$\neg(p \vee q)$	$\neg p \wedge \neg q$	$\neg(p \vee q) \Rightarrow \neg p \wedge \neg q$
T	T	F	F				
T	F	F	T				
F	T	T	F				
F	F	T	T				

ii Is the above result a tautology, a logical contradiction, or neither?

b Consider the two propositions: p: I work hard.
                q: I get a promotion.

i Write in words, the conjunction of p and q.

ii Write in words, the inverse of $p \Rightarrow q$.

c At the TULCO telecommunications company, the probability of a worker working hard is 0.8. Two workers are selected at random.

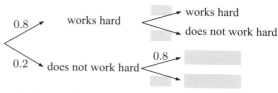

i Copy and complete the tree diagram.

ii Use the tree diagram to find the probability that for two randomly selected workers:

(1) both work hard   **(2)** only one works hard.

8 The Venn diagram below shows the number of students out of a class of 30 who like Rock music (R), Jazz music (J), and Classical music (C).

17 like Rock music,
9 like Jazz, and
7 like Classical music.

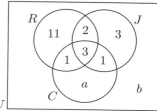

a **i** Find the number of students a who like only Classical music.

ii Describe in words the region containing 11 students.

iii Find the number of students b.

iv Describe region b in words.

b Sketch the Venn diagram and shade in the region $R' \cap J$.

c A student is chosen at random from this class. Determine the probability that the student likes:

i all three types of music   **ii** only Classical music

iii Rock music, given that they like Jazz music.

d Two students are randomly selected from the class. Determine the probability that:

i both students like Rock music only

ii one student likes Rock only, and the other likes all three types.

9 Let p, q, and r represent the propositions:

   p: I play computer games. q: I get a headache.
   r: I go to sleep.

a In words, write down the following statements:

i $p \Rightarrow r$ **ii** $\neg p \Rightarrow \neg r$

b Out of the choices *inverse*, *converse*, or *contrapositive*, which word best describes the relationship between $p \Rightarrow r$ and $\neg p \Rightarrow \neg r$?

c Write, in symbolic form, the statement:

"*If I play computer games then I get a headache and I go to sleep*".

d Complete the truth table below for the compound proposition $(p \Rightarrow (q \wedge r)) \wedge p \Rightarrow r$.

p	q	r	$q \wedge r$	$p \Rightarrow (q \wedge r)$	$(p \Rightarrow (q \wedge r)) \wedge p$	$(p \Rightarrow (q \wedge r))$ $\wedge p \Rightarrow r$
T	T	T				
T	T	F				
T	F	T				
T	F	F				
F	T	T				
F	T	F				
F	F	T				
F	F	F				

e State whether $(p \Rightarrow (q \wedge r)) \wedge p \Rightarrow r$ is a tautology, a logical contradiction, or neither.

10 A representative survey of students' weekly spending at the school canteen produced the following results:

Weekly expenditure ($)	Frequency
0 - 9.99	11
10 - 19.99	25
20 - 29.99	34
30 - 39.99	30
40 - 49.99	20

a **i** Write down the number of students in the survey.
  ii Find the probability that a student spends less than $20 per week.
  iii If there are 1500 students at the school, estimate the number of students who spend between $30 and $50 each week.

b One student is chosen at random from those surveyed. Given that the student spends more than $10 each week, find the probability that the student spends more than $30 each week.

c Three students are selected at random from those surveyed. Find the probability that:
  i all 3 students spend more than $40 each week
  ii none of the students spend more than $40 each week
  iii exactly one of the students spends more than $40 each week
  iv at least 2 of the students spend more than $40 each week.

11 80 students were asked what type of television programmes they had watched the previous evening.
  35 watched sport (S).
  42 watched news (N).
  50 watched drama (D).
  10 watched all three types.
  7 watched sport and news only.
  12 watched news and drama only.
  14 watched sport and drama only.

a Draw a Venn diagram to display this information.
b Determine the number of students who watch neither sport nor drama nor news.
c A student is selected at random. Determine the probability that the student:
  i watched only drama
  ii watched sport, given that they watched news.
d Let p and q be the propositions:
  p: You watch sport on television. q: You like sport.
  i Write in words the proposition $p \Rightarrow q$.
  ii The converse of the proposition $p \Rightarrow q$ is $q \Rightarrow p$. Write this proposition in words.
  iii Consider the following proposition:
    "If you do not like sport then you do not watch sport on television."
    Express this statement using symbols only.
  iv Is the proposition in **iii** the inverse, the converse, or the contrapositive of the proposition in **i**?

12 The probability of rain falling on any day in Dunedin is 0.4. The tree diagram below shows the possible outcomes when two consecutive days are considered.

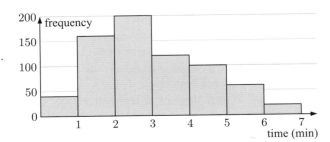

a Copy and complete the tree diagram by filling in the boxes.
b Use the tree diagram to determine the probability of:
  i rain on both days
  ii no rain on one day.
c Find the probability of the weather being fine for five consecutive days.
d Consider the propositions:
  p: It is raining. q: I wear my raincoat.
  Write the following propositions using words only:
  i $p \Rightarrow q$
  ii $\neg q \Rightarrow \neg p$
e Consider the compound proposition $\neg(p \wedge q) \Rightarrow \neg p$.
  i Construct a truth table for this proposition.
  ii Give an example for when this proposition is false.

13 The frequency histogram below illustrates the time taken for a group of patrons to locate their seats in a large concert hall.

a How many patrons were surveyed at this event?
b If a patron is randomly selected at this event, determine the probability that he or she:
  i was seated within 4 minutes
  ii was not seated within 1 minute
  iii took between 1 and 4 minutes to be seated.
c The concert hall consists of 40 rows of seats. The first row contains 22 seats, and each subsequent row contains 4 more seats than the previous row.
  i If u_n represents the number of seats in the nth row, show that $u_n = 4n + 18$.
  ii Calculate the total number of seats in the hall.
d Tickets for an event at the concert hall are randomly allocated. Determine the probability that a patron is allocated a ticket for a seat:
  i in the back row
  ii within the first 10 rows.

14 **a** Two marbles are selected without replacement from bag A containing 8 red and 5 blue marbles.
  i Draw a tree diagram to represent the sample space.
  ii Use the tree diagram to determine the probability of obtaining:
    (1) two marbles that are the same colour
    (2) at least one blue marble.

b A second bag B contains 7 red and 4 blue marbles. A five sided spinner with the numbers 1, 2, 3, 4, 5 on it is used to select from which bag the two marbles are taken. If an even number is spun then bag A is chosen, and if an odd number is spun then bag B is chosen.

   i Draw a new tree diagram to represent all the possibilities.

   ii Use the tree diagram to determine the probability of obtaining:

     (1) an even number on the spinner

     (2) two blue marbles.

15 a A box of chocolates contains 4 hard and 7 soft centres. One chocolate is selected randomly and eaten. If it has a hard centre, then a second chocolate is selected randomly and eaten.

The tree diagram below represents the possible outcomes for this event.

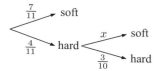

   i Find the probability that two chocolates are eaten.

   ii Find the value of x.

   iii Find the probability that two chocolates with hard centres are eaten.

   iv Find the probability that one chocolate of each type is eaten.

b A second box of chocolates contains all soft centres flavoured either with strawberry or cherry. There are 5 strawberry and 7 cherry flavoured chocolates. One chocolate is selected at random and eaten, then another is selected.

   i Draw a tree diagram to represent all possible outcomes.

   ii Use your tree diagram to determine the probability that:

     (1) both chocolates are strawberry

     (2) the second chocolate is strawberry.

   iii A girl who does not like cherry flavoured chocolates randomly selects a chocolate from a new identical box, tries it, and discards it if it is cherry flavoured. She then selects another, repeating the process until a strawberry flavoured chocolate is found.

Determine the probability that the girl selects a total of 4 chocolates.

TOPIC 4: STATISTICAL APPLICATIONS

EXPECTATION

If there are n trials of an experiment, and an event has probability p of occurring in each of the trials, then the number of times we **expect** the event to occur is np.

For an experiment with outcomes $x_1, x_2, ..., x_n$ and associated probabilities $p_1, p_2, ..., p_n$, the **expectation** from the experiment is $\sum_{i=1}^{n} x_i p_i$.

A game is **fair** if the expected outcome from the game is zero.

THE NORMAL DISTRIBUTION

A **continuous random variable** is a variable which can take any real value within a certain range.

For a continuous variable X, the probability that X is *exactly* equal to a particular value is zero. It only makes sense to talk about the probability that X lies in a certain **interval**.

The **normal distribution** occurs frequently in nature. It is symmetric about its mean and has a bell-shaped curve. The area under the curve is 1.

A random variable X which is normally distributed about mean μ with standard deviation σ, is denoted $X \sim N(\mu, \sigma^2)$.

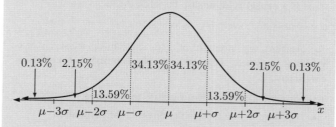

For problems involving the normal distribution, always draw a diagram.

You should be able to find the probability that X lies in a given interval of a known distribution.

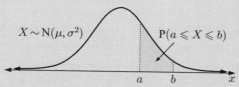

Suppose $X \sim N(\mu, \sigma^2)$ and that $P(X \leqslant k) = p$. We say that k is the **quantile** corresponding to probability p. We calculate k using the **inverse normal function**.

TWO VARIABLE STATISTICS

Correlation refers to the relationship or association between two variables.

We can use a **scatter diagram** of the data to help identify **outliers** and to describe the correlation. We consider **direction**, **strength**, and whether a relationship is **linear**.

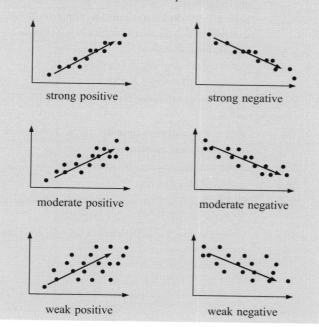

strong positive

strong negative

moderate positive

moderate negative

weak positive

weak negative

These points are roughly linear. These points do not follow a linear trend.

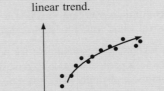

An **outlier** is an isolated point which does not follow the general trend of the data. An outlier should only be included if it is proved to be genuine data.

If a change in one variable *causes* a change in the other, we say there is a **causal relationship** between them.

Correlation does *not* imply causation.

To measure the strength of the relationship between two variables, we use **Pearson's product-moment correlation coefficient**, r.

The correlation coefficient always lies in the range $-1 \leqslant r \leqslant 1$.

- The sign of r indicates the direction of correlation.
- The size of r indicates the strength of correlation.

Value	Strength of correlation
$|r| = 1$	perfect correlation
$0.95 \leqslant |r| < 1$	very strong correlation
$0.87 \leqslant |r| < 0.95$	strong correlation
$0.5 \leqslant |r| < 0.87$	moderate correlation
$0.1 \leqslant |r| < 0.5$	weak correlation
$0 \leqslant |r| < 0.1$	no correlation

The straight **line of best fit** drawn by eye should pass through the **mean point** $(\overline{x}, \overline{y})$.

You should be able to use a calculator to find the **least squares regression line** which best fits a set of bivariate data.

When using a line of best fit to estimate values, **interpolation** is usually reliable whereas **extrapolation** may not be.

The χ^2 test for independence

Step 1: State H_0 called the **null hypothesis**. This is a statement that the two variables being considered are independent.
State H_1 called the **alternative hypothesis**. This is a statement that the two variables being considered are not independent.

Step 2: Calculate the **degrees of freedom**
$$df = (r - 1)(c - 1).$$

Step 3: Quote the **significance level** required, 10%, 5%, or 1%.

Step 4: State the **rejection inequality** $\chi^2_{calc} > k$ where k is the **critical value** of χ^2.

Step 5: Construct the expected frequency table.

Step 6: Use technology to find χ_{calc}.

Step 7: We either reject H_0 or do not reject H_0, depending on the result of the rejection inequality.

Step 8: We could also use a **p-value** to help us with our decision making.
For example, at a 5% significance level:
If $p < 0.05$, we reject H_0.
If $p > 0.05$, we do not reject H_0.

SKILL BUILDER - SHORT QUESTIONS

1 The probability of Roger getting a 'first serve' in, in tennis, is $\frac{7}{9}$. How many 'first serves' would you expect Roger to get in out of 180 attempts?

2 Consider the continuous random variable $X \sim N(44, 20)$. Suppose $P(X \leqslant m) = 0.65$ and $P(m \leqslant X \leqslant n) = 0.2$. Find $m - n$.

3 A normally distributed random variable X has mean 120 and standard deviation 25.
a Find $P(135 \leqslant X \leqslant 145)$.
b Find $P(X \geqslant 90)$.
c Find k given that $P(X \geqslant k) = 0.175$.

4 **a** If 3 coins are tossed, find the probability that two fall heads and the other falls tails.
b Suppose 3 coins are tossed 400 times. On how many occasions would you expect to see exactly one tail?

5 The scores obtained by students in an IB examination are normally distributed with mean 60 and standard deviation 12.
a Find the probability that a randomly chosen candidate scored more than 80.
b A group of 90 students from one school sat this examination. How many are likely to have scored between 50 and 70?
c Estimate the minimum score for a student to be in the top 30% of candidates.

6 To test the effects of adrenaline on mice, subjects were timed through an obstacle course with and without adrenaline injections. The results in seconds are shown below:

Subject	With adrenaline (x)	Without adrenaline (y)
A	35	38
B	42	44
C	31	32
D	50	38
E	62	70
F	28	31
G	57	68
H	39	49

a Determine the equation of the regression line of y against x.
b State Pearson's correlation coefficient.
c Comment on the strength of the correlation.
d Which data point is an outlier?
e Using your regression line, estimate the benefit of an adrenaline injection for a mouse that takes 48 seconds without it.

7 Consider the contingency table shown.

	Y_1	Y_2
X_1	32	14
X_2	25	19

a Construct a corresponding table of expected values, giving each result as a whole number.
b Find χ^2_{calc} for this data.

8 The times taken for the 200 runners in the school cross country event were normally distributed with mean 26 minutes and standard deviation 4 minutes.
a Estimate the number of runners who took:
i less than 22 minutes to complete the course
ii more than 27 minutes to complete the course.

b The fastest 40% of runners finished quicker than what time?

9 When Michael shoots for goal in basketball, he scores either 0, 1, 2, or 3 points, with the probabilities shown.

Number of points	0	1	2	3
Probability	0.19	0.34	0.38	0.09

On average, how many points do you expect Michael to score with each shot?

10 The following table lists the ages of contestants in a game and the times they take to complete the task.

Age (x years)	28	40	21	38	30	26
Time (y min)	20	32	15	40	26	25

Age (x years)	18	32	25	29	20	24
Time (y min)	19	28	21	25	16	22

a Find the value of the correlation coefficient r.

b State the meaning of the value found in **a**.

c Write down the equation of the linear regression line in the form $y = mx + c$.

d Explain the meaning of the coefficient m in the equation in **c**.

11 A survey of people found the following preferences for flavoured milk:

	Chocolate	Coffee	Strawberry	Caramel
Adult	26	30	15	12
Child	40	12	15	10

It is claimed that the preferred flavour is independent of age.

a Write suitable null and alternative hypotheses for a χ^2 test.

b Find the χ^2 statistic.

c Given $\chi^2_{crit} = 7.81$, is there evidence at a 5% level to support the claim?

12 The table shows the exchange rate for Argentian pesos against USD, and interest rates in Argentina at the corresponding times.

Exchange rate	2.85	2.95	2.90	2.75	2.65
Interest rate	7.40	7.50	7.55	7.25	7.25

Exchange rate	2.80	3.05	2.98	2.95
Interest rate	7.35	7.65	7.75	7.60

a Draw a scatter diagram for the given data.

b Find the correlation coefficient between the two variables.

c Describe the nature and strength of the relationship between the variables over this period of time.

13 Let $X \sim N(17, 3^2)$.

a State the mean and standard deviation for X.

b Find $P(X > 21)$.

c Given that $P(X < k) = P(X > 22)$, determine the value of k.

14 A nursery has developed a new hybrid plant. They claim that this hybrid will grow equally well in any light conditions. They have provided the following data to support their claim.

	Height < 60 cm	Height ⩾ 60 cm
Sunlight	37	43
Shade	22	18
Dark	25	19

a Write suitable null and alternate hypotheses for a χ^2 test.

b Find the χ^2 statistic for the plant data.

c Given $\chi^2_{crit} = 5.991$, is there evidence at a 5% level to support the nursery's claim?

15 Consider the following data on farm production.

Monthly rainfall (mm)	5	10	15	20	25	30
Yield (tonnes)	14	21	29	31	30	28

a Determine the coefficient of correlation r.

b What does the value of r suggest about the nature and strength of the relationship between monthly rainfall and crop yield?

c Draw the scatter diagram for the farm data.

d Explain why the coefficient of correlation may not be an appropriate measure for this data.

16 The mean birth weight of babies in a population is normally distributed with mean 3.4 kg and standard deviation 300 grams.

a What proportion of babies in this population have birth weight:

  i in excess of 4 kg

  ii between 3 and 4 kilograms?

b A *low birth weight* corresponds to any newborn weighing in the lowest 10% of birth weights. State the weight below which a baby is classified as having a *low birth weight*.

17 A random variable X is normally distributed with mean 12 and standard deviation 3.

a Represent this information on a diagram like the one shown. Include an appropriate scale on the horizontal axis.

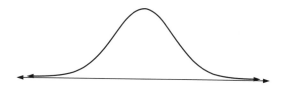

b Calculate $P(10 \leqslant X \leqslant 14)$.

c Suppose $P(X > k) = 0.3$.

  i Show the approximate position of k on your diagram from **a**.

  ii Find k.

18 A company produces LED globes for household use. It is known that 0.5% of the LED globes are faulty.

a As a percentage, calculate the probability that:

  i an LED globe is not faulty

  ii two consecutive randomly chosen LED globes are faulty.

b If the company sells 3000 LED globes to a retailer, how many would they expect to be faulty?

19 Consider $X \sim N(p, 1)$.

   a On a diagram like the one shown, sketch the region corresponding to $P(X > p + 1)$.

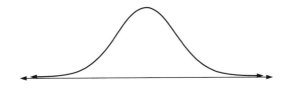

   b Given that $P(X < k) = P(X > p + 1)$, write down the value of k.

20 Test the independence of the following factors at the 5% level of significance, given $\chi^2_{crit} = 9.488$.

	Factor Y_a	Factor Y_b	Factor Y_c
Factor X_a	6	3	7
Factor X_b	21	35	28
Factor X_c	16	11	22

For your test, clearly define the null and alternate hypotheses, the number of degrees of freedom, the calculated χ^2 value, and the p-value. Clearly state a conclusion for your test.

21 In a sample of 2.5 kg bags, the number of new potatoes in each bag and their median weight were recorded:

Median weight (x g)	89	97	105	110	125	140	145	150
Number in bag (y)	28	26	25	23	21	18	18	16

   a Determine the equation of linear regression for this data.

   b Use your equation to estimate the number of potatoes in a bag if the median weight is:

     i 100 grams   **ii** 200 grams.

   c Which of the answers in **b** is likely to be more reliable? Explain your answer.

22 Blake plays hockey, and scores at least one goal in 15% of the games he plays. Miles plays football, and scores at least one goal in 25% of his games. During the winter sporting season, the boys play one game of their chosen sports each weekend.

   a For a single weekend during the winter sporting season, determine the probability that:

     i both boys score a goal

     ii at least one boy scores a goal.

   b There are 15 games played over the winter sporting season. Find the expected number of weekends when neither boy scores a goal.

23 A bag contains 1 blue ticket, 3 red tickets, and 8 yellow tickets. The reward for a blue ticket is $40, for a red ticket is $20, and for a yellow ticket is $5. Tickets are selected randomly and one at a time, with replacement.

   a Calculate the expected return for one trial of this game.

   b If tickets cost $15 per game, explain why it would not be advisable to play this game.

   c Find the number of extra red tickets that should be added to the bag, to make the game fair.

24 9 students sat a mathematics examination. The results that they obtained and the number of hours that each of them studied are shown in the following table.

Study time (h)	7	6	3	16	15	11	18	32	20
Result (%)	56	42	25	80	65	60	85	96	90

   a Write down the equation of the straight line of best fit.

   b Tony's score in the examination was 70%. Using the line of best fit, estimate how long he studied for.

   c In terms of marks obtained in the examination, interpret the y-intercept and the gradient of the equation of the line of best fit.

25 Katie has a theory that the satisfaction of employees in summer is dependent on the quantity of ice cream their boss supplies. She collects the following data:

		satisfaction of employees		
		high	medium	low
quantity	high	10	4	4
of	low	14	20	10
ice cream	zero	64	50	54

   a Calculate the table of expected values assuming the variables are independent.

   b Given $\chi^2_{crit} = 13.28$, conduct a χ^2 test at the 1% level to test the validity of Katie's claim. State clearly a conclusion for your test.

26 The time taken for a skier to complete a particular downhill run is normally distributed with mean 45 seconds and standard deviation 4 seconds.

   a Find the probability that the skier completes:

     i one downhill run in under 40 seconds

     ii two consecutive downhill runs in under 40 seconds each.

   b If the skier completes a total of 60 independent runs, how many times would you expect the run to take between 44 and 47 seconds?

27 The average height h (in mm) of grass t days after being mowed, is shown in the table below.

Days since mowed, t	0	1	2	3	4	5	6	7	8	9
Height, h (in mm)	5	5.7	5.7	6.2	6.8	7.1	8	8.3	9	9.3

   a Calculate the value of Pearson's product moment-correlation coefficient r.

   b Explain the significance of the size and sign of r.

   c The regression line for h on t is $h \approx 0.4879t + 4.9145$. Using this equation, estimate the:

     i height of the grass after 14 days

     ii time required for the height of the grass to reach 20 mm.

28 Students in an art course were asked to select one specialisation: painting, sketching, or sculpting. The number of students who chose each activity is shown below.

	Painting	Sketching	Sculpting
Male	30	35	15
Female	20	15	25

A χ^2 test is to be used to determine whether gender is independent of the specialisation chosen. The χ^2 test is performed at the 1% significance level, with corresponding critical χ^2 value 9.210.

a Assuming gender is independent to the specialisation chosen, find the expected number of male sculptors.

b Calculate the value of the χ^2 test statistic.

c Perform the test, and state clearly its conclusion.

SKILL BUILDER - LONG QUESTIONS

1 The volume of drink dispensed by a coffee machine is normally distributed with mean 254 mL and standard deviation 2.3 mL.

  a Find the probability that a randomly selected cup of coffee from the machine will have volume less than 254 mL.

  b Find the percentage of cups of coffee dispensed by the machine, which have volume between 252 and 256 mL.

  c A sample of 80 cups of coffee is taken from the machine. Determine the number of drinks which will be expected to have volume at least two standard deviations above the mean.

  d The machine operator guarantees that at least 95% of cups of coffee will have volume at least 250 mL.

    i Is the guarantee valid?

    ii A technician adjusts the machine so the standard deviation is now 2.5 mL. What effect does this have on the operator's guarantee?

2 An investigation into the weight of packed vegetables found the following:

Number of tomatoes in a 1.5 kg bag	Median weight of tomatoes in bag (g)
15	90
11	125
14	110
12	125
14	136
17	82
12	115
10	150

  a On graph paper, draw a scatter diagram to illustrate this data. Use 1 cm on the horizontal axis to represent 1 tomato, and 1 cm on the vertical axis to represent 20 grams.

  b Draw a suitable line of best fit by hand.

  c Describe the apparent relationship between the number of tomatoes in a bag and the median weight.

  d Use your graph to estimate:

    i the median mass when there are 13 tomatoes in a bag

    ii the number of tomatoes in a 1.5 kg bag if the median weight of those tomatoes is 100 grams.

  e Enter the data into your calculator.

    i Write down the equation of the linear regression line.

    ii Hence estimate the median weight of tomatoes when there are 20 in a 1.5 kg bag.

    iii Comment on the reliability of your estimate.

    iv Write down the value of the correlation coefficient r.

3 An experiment was conducted in an orchard to determine whether a new type of fertiliser increased the yield of three varieties of oranges. The number of oranges per tree was calculated and the results were as follows:

	Variety A	Variety B	Variety C	Totals
New fertiliser	65	48	75	188
Old fertiliser	40	54	58	152
Totals	105	102	133	340

The table below shows some of the expected yields of each type of orange assuming the yield is independent of the fertiliser.

	Variety A	Variety B	Variety C
New fertiliser	a	c	73.5
Old fertiliser	b	45.6	59.5

  a Write down a suitable null hypothesis H_0 and alternate hypothesis H_1 to test this independence.

  b Show that the expected yield of Variety A oranges with new fertiliser is 58.1.

  c Write down the χ^2 value for the above data.

  d Find the number of degrees of freedom.

  e Find the p-value.

  f At the 5% significance level, what conclusion can be drawn regarding the use of the new fertiliser?

4 A jeweller measured the volume and mass of some samples of silver he had been given. He suspected that one of the samples might be a fake. The results are listed in the table.

Sample	A	B	C	D	E	F
Volume (x cm^3)	3	6	4	7	16	8
Mass (y g)	40	95	50	160	285	130

Sample	G	H	I	J	K	L
Volume (x cm^3)	5	12	9	6	10	11
Mass (y g)	65	210	155	90	170	190

  a Draw a scatter diagram for this data. Use 1 cm to represent 5 cm^3 on the horizontal axis, and 1 cm to represent 100 g on the vertical axis.

  b **i** Calculate the mean for both volume ($\overline{x}$) and mass ($\overline{y}$).

    ii Draw the straight line of best fit which should pass through the point ($\overline{x}, \overline{y}$).

  c Write down the linear coefficient of correlation r.

  d Describe the relationship which appears to exist between the volume and mass of the samples of silver.

  e Do you agree with the jeweller that there is a fake sample?

  f **i** Remove the suspect value from the data and then find the equation of the linear regression line for the remaining data.

    ii Use your equation to find the expected mass of the sample of silver with the same volume as the suspect sample.

    iii Calculate the percentage error between the given and expected masses of the suspect sample, based on the expected mass.

5 Fifty words were randomly selected from story A and another fifty words were randomly selected from story B. The number of letters in each word were recorded and the results are summarised in the following frequency table.

Number of letters in a word	Frequency of words in story A	Frequency of words in story B
2	5	7
3	13	9
4	6	10
5	8	6
6	5	6
7	5	8
8	2	2
9	4	1
10	2	1

a Draw frequency histograms for both stories on equivalent axes.

b Determine the five-number summary for the words in each story, and display the statistics as side-by-side box and whisker plots.

c Write down the mean and standard deviation for the words in each story.

d Construct a 2×2 contingency table by adding the number of words with less than 5 letters and the number of words with 5 or more letters for each story.

e Test for independence at the 10% significance level for the chi-square distribution, given $\chi^2_{crit} = 2.706$.

f What conclusion can be drawn from the test?

g Is there any evidence to support the claim that the two stories were written by different authors?

6 Jack believes that students' choice of breakfast cereal is related to gender. He conducts a survey at his school and records the following information. He plans to use a χ^2 test to check if his belief is correct.

	Muesli	Rolled Oats	Corn Flakes	Weetbix
Female	4	14	48	24
Male	8	30	26	26

a Write a suitable null hypothesis for the χ^2 test.

b Write down the expected values for this data.

c Find the χ^2 statistic.

d Given that $\chi^2_{crit} = 7.815$ for a test at the 5% significance level, determine whether or not to reject H_0.

e What conclusion can Jack draw from the test?

7 The scatter diagram shows the age and annual income for 10 randomly chosen individuals.

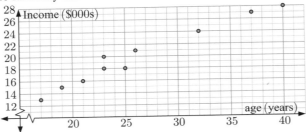

a Construct a table to display the data in the graph.

b Determine the value of the correlation coefficient r.

c Describe the relationship between the age and annual income for these individuals.

d Determine the equation of the linear regression line for age and income.

e Hence estimate the annual income for someone who is:
  i 30 years old   **ii** 60 years old.

f Comment on the reliability of your answers in **e**.

8 Two four-sided tetrahedral dice are tossed simultaneously. The faces of each die are labelled 1, 2, 3, and 4. Let S represent the sum of the results from both dice.

a The grid below shows the values of S for each possible set of results. Copy the grid and fill in the missing values.

Die 2

		1	2	3	4
	1	2	3	4	5
	2	3	4	5	
Die 1	3	4	5		
	4	5			

b Find:
  i $P(S = 5)$ **ii** $P(S > 5)$ **iii** $P(S = 5 \mid S > 3)$

c A game is constructed whereby points are awarded according to the value of S.

Value of S	$S = 2$	$3 \leqslant S \leqslant 5$	$S > 5$
Number of points won or lost	32	16	−8

  i Determine the expected number of points for each roll of the dice.

  ii The number of points lost for $S > 5$ is to be altered so that the expectation for the game is zero. How many points must be lost for $S > 5$?

9 To test the claim that age and reaction time are independent, a researcher collected the following data from a group of 200 individuals.

	Age				
Reaction time	Child	Teen	Adult	Senior	Total
Fast	15	25	19	9	68
Medium	9	21	25	11	66
Slow	5	18	22	21	66
Total	29	64	66	41	200

A χ^2 test at the 5% level of significance is performed to test the claim.

a For this test, write down the:
  i null hypothesis   **ii** alternative hypothesis.

b Of the 200 people in this sample, how many:
  i were teenagers   **ii** had a slow reaction time?

c If we assume age is independent of reaction time, what is the expected number of teenagers with a slow reaction time? Do not round your result.

d Write down the number of degrees of freedom.

e Calculate the value of the test statistic χ^2_{calc}.

f What conclusion would the researcher give based on this χ^2 test, given $\chi^2_{crit} = 9.49$?

g If the χ^2 test were carried out at the 1% level of significance, with $\chi^2_{crit} = 16.81$, would it lead to the same conclusion? Explain your answer.

TOPIC 5: GEOMETRY AND TRIGONOMETRY

PYTHAGORAS' THEOREM

In a right angled triangle with legs a and b, and hypotenuse c,

$$a^2 + b^2 = c^2$$

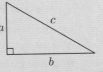

The *converse* of Pythagoras' theorem can be used to establish the presence of a right angle.

THE NUMBER PLANE

- also called the Cartesian plane
- consists of the set of all points (x, y) relative to an origin O.

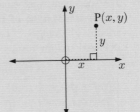

DISTANCE BETWEEN TWO POINTS

The distance between A(x_1, y_1) and B(x_2, y_2) is

$$AB = \sqrt{(x_2 - x_1)^2 + (y_2 - y_1)^2}.$$

THE MIDPOINT FORMULA

The midpoint M half-way between A(x_1, y_1) and B(x_2, y_2) has coordinates M$\left(\dfrac{x_1 + x_2}{2}, \dfrac{y_1 + y_2}{2}\right)$.

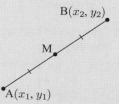

GRADIENT

The **gradient** of a line is a measure of its steepness.

The gradient of the line AB passing through A(x_1, y_1) and B(x_2, y_2) is $m = \dfrac{y\text{-step}}{x\text{-step}} = \dfrac{y_2 - y_1}{x_2 - x_1}$.

The gradient of any horizontal line is zero.

The gradient of any vertical line is undefined.

The gradients of parallel lines are equal. $m_1 = m_2$

The gradients of perpendicular lines are negative reciprocals. $m_1 = -\dfrac{1}{m_2}$

Collinear points are points which lie on the same straight line.

EQUATION OF A LINE

The equation of a line can be presented in:

- **gradient-intercept form** $y = mx + c$ where m is the gradient and c is the y-intercept.
- **general form** $ax + by + d = 0$

You should be able to find the equation of a line given:

- its gradient and the coordinates of any point on the line
- the coordinates of two distinct points on the line.

The straight line with gradient m which passes through (x_1, y_1) has equation $\dfrac{y - y_1}{x - x_1} = m$.

Two straight lines may:

- intersect in a single point
- be parallel and never meet
- be coincident and have infinitely many points of intersection.

We find where two lines meet by finding their simultaneous solution. You should be able to do this:

- graphically
- using technology.

PERPENDICULAR BISECTORS

The perpendicular bisector of AB is the set of all points which are the same distance from A and B.

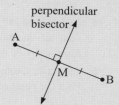

To find the perpendicular bisector of line segment AB:

- find the midpoint M of AB
- find the gradient m of AB
- the perpendicular bisector has gradient $-\dfrac{1}{m}$ and passes through M.

GEOMETRY OF 2-DIMENSIONAL FIGURES

The **perimeter** of a figure is the distance around its boundary.

The perimeter of a circle is called its **circumference**.

The **area** of a closed figure is the number of square units it contains.

You must be able to:

- convert between units of length and area
- calculate the perimeter and area of figures
- solve problems involving perimeter and area.

GEOMETRY OF 3-DIMENSIONAL FIGURES

The **surface area** of a three-dimensional figure with plane faces is the sum of the areas of the faces.

The **volume** of a solid is the amount of space it occupies.

The **capacity** of a container is the quantity of fluid it is capable of holding. You should understand how the units of volume and capacity are related.

You should be able to calculate the surface area and volume of different 3-dimensional figures, including solids of uniform cross-section, pyramids, spheres, and cones.

RIGHT ANGLED TRIANGLE TRIGONOMETRY

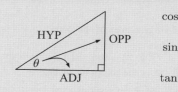

$$\cos \theta = \frac{\text{ADJ}}{\text{HYP}}$$

$$\sin \theta = \frac{\text{OPP}}{\text{HYP}}$$

$$\tan \theta = \frac{\text{OPP}}{\text{ADJ}}$$

You should be able to use these ratios to find unknown sides and angles of right angled triangles. These skills can be applied to both 2-dimensional and 3-dimensional objects. For example, you should be able to find the angle between a line and a plane.

AREA OF A TRIANGLE

The area of any triangle is a half of the product of two sides and the sine of the included angle between them.

$$\text{Area} = \tfrac{1}{2}ab \sin C$$

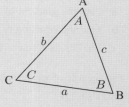

THE COSINE RULE

$$a^2 = b^2 + c^2 - 2bc \cos A$$

$$\text{or} \quad \cos A = \frac{b^2 + c^2 - a^2}{2bc}$$

Use the cosine rule when given:

- two sides and an included angle
- three sides.

THE SINE RULE

$$\frac{a}{\sin A} = \frac{b}{\sin B} = \frac{c}{\sin C}$$

Use the sine rule when given:

- one side and two angles
- two sides and a non-included angle.

SKILL BUILDER - SHORT QUESTIONS

1 a Plot the points S(2, 1) and T(−4, −2) on coordinate axes.

b Find the length of the line segment from S to T.

c The point R(b, 4) is 5 units from S, where $b > 0$. Find the value of b.

2 a Calculate the gradient of the line passing through G(3, 1) and H(−3, 3).

b Find the midpoint of GH.

c Determine the equation of the perpendicular bisector of GH.

3 Consider the straight lines $L_1: 4x − 3y = 7$ and $L_2: 2x + ky = 5$.

a Find the gradient of L_1.

b Find k if the two lines L_1 and L_2 are:
  i parallel   **ii** perpendicular.

4 A solid right-circular cone has base radius 12 cm and vertical height 15 cm.

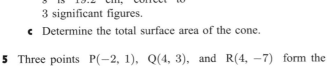

a Determine the angle between the slanting side and the base.

b Show that the slant height s is 19.2 cm, correct to 3 significant figures.

c Determine the total surface area of the cone.

5 Three points P(−2, 1), Q(4, 3), and R(4, −7) form the vertices of triangle PQR. M is the midpoint of PQ.

a Plot the points P, Q, and R on a set of axes.

b Write down the coordinates of M.

c A circle with centre R is drawn passing through M. Calculate the area of this circle, giving your answer correct to the nearest whole number.

6 A line passes through the points (−3, 4) and (−1, 10).

a Find the gradient of the line.

b Find the equation of the line.

c Find its axes intercepts.

7 The vertices of an isosceles triangle are A(−3, 4), B(−$\frac{1}{2}$, −1), and C(5, −2).

a Find the coordinates of the midpoint of the longest side.

b Calculate the length of the line joining the midpoint of the longest side to the opposite vertex.

c Given that the longest side measures 10 units, find the size of the equal angles of the triangle.

8 The straight line L_1 has equation $3x + 2y = 5$.

a Find the gradient of the line L_2 which is perpendicular to L_1.

b Determine the equation of the line L_2 given that it passes through the point (2, −1).

c Find the x-intercept of L_2.

9 An 8 m long ladder is resting against the wall of a house. The angle that the ladder makes with the wall is 35°.

a Calculate the distance from the foot of the ladder to the base of the wall.

b Suddenly, the ladder slips so that its foot moves 2 m further away from the base of the wall. Calculate the angle that the ladder now makes with the horizontal.

10 A triangle ABC has vertices A(−2, 2), B(4, 4), and C(5, 1).

a Plot the points A, B, and C on coordinate axes.

b Using gradients, show that the triangle is right angled at B.

c Determine the length of side AC.

11 Two parallel lines are shown in the diagram.

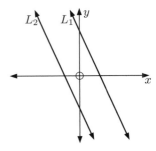

a L_1 has equation $7x + 3y = 12$. Find the gradient of L_2.

b Find the equation of the line perpendicular to L_2, which passes through (7, 1).

12 a The straight line $2x − 4y = 7$ meets the x-axis at the point (k, 0). Find the value of k.

b The straight line $ax + by + d = 0$ passes through the points (3, 2) and (−6, 0). Find the smallest possible integer values of a, b, and d.

13 Triangle ABC is acute with AC = 12 cm and BC = 11 cm. The area of triangle ABC is 33 cm².

a Sketch triangle ABC, showing all of the information provided.

b Show that $A\widehat{C}B = 30°$.

c Hence, calculate the length of AB.

14 Line L has equation $y = 3 − 2x$.

a If the point P(3, k) lies on line L, determine the value of k.

b Write down the gradient of line L.

c Hence, find the equation of the line perpendicular to L, which also passes through point P.

15 Triangle ABC is right angled at C. AB is 5 cm, and BC is 2 cm.

a Sketch triangle ABC, clearly showing all of the information provided.

b Calculate $B\widehat{A}C$ correct to 4 significant figures.

c Find the length of AC correct to 3 significant figures.

16

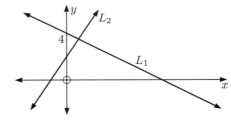

a L_1 has gradient −$\frac{1}{2}$ and passes through (0, 4). Find the equation of L_1, expressing your answer in the form $ax + by + d = 0$ where a, b, $d \in \mathbb{Z}$.

b L_2 passes through (−2, −1) and (4, 8). Find the point of intersection of L_1 and L_2.

17 a Find k if $kx + 3y = 7$ has x-intercept 4.

b The two straight lines $ax − y = 4h$ and $ax + 2y = h$ intersect at the point (1, −1). Find the values of a and h.

18 Consider the graphs alongside:

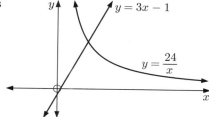

a Find the coordinates of the point shown where the graphs intersect.

b Find the equation of the line perpendicular to $y = 3x - 1$ which passes through the point of intersection found in **a**.

19 Triangle ABC is drawn inside a semi-circle of radius 5 cm as shown. Side AB has length 8 cm and side BC has length 6 cm.

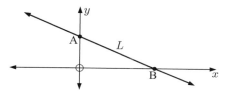

a Use the theorem of Pythagoras to show that triangle ABC is right angled at B.

b Find the shaded area.

20 Line L has equation $3x + 2y - 6 = 0$ and passes through the points A and B as shown.

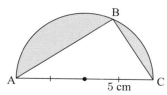

a Find the coordinates of the points A and B.

b Determine the gradient of line L.

c Line M passes through A and is perpendicular to line L. Write down the equation of line M.

21 ABC is an equilateral triangle with sides 10 cm long. P is a point within the triangle. P is 5 cm from A and 6 cm from B.

a Draw a diagram clearly showing all of the information provided.

b Calculate: **i** $B\widehat{A}P$ **ii** $C\widehat{A}P$

c Hence, find the length of CP.

22 Triangle ABC is right angled at B, and BC = 9 cm. Point P lies on AC and is equidistant from A and B.
Angle CPB = 102°.

a Draw a diagram showing the given information.

b Find the length of AC.

23 At 2:35 pm Fari sees an airplane directly overhead. At 2:38 pm he estimates that the angle of elevation to the plane is 25°.
If the plane is travelling in a straight line at 110 m s^{-1}, calculate:

a the height of the plane above the ground

b the angle of elevation of the plane at 2:42 pm.

24 In triangle ABC, AB = 72 cm, BC = 61 cm, and $A\widehat{B}C = 43°$.

a Calculate the length of AC.

b Find the measure of $A\widehat{C}B$.

25 The diagram shows a vertical pole PQ, which is supported by two wires fixed to the horizontal ground at A and B. $P\widehat{B}Q = 36°$, $B\widehat{A}Q = 70°$, $A\widehat{B}Q = 30°$, and the distance BQ is 40 m.

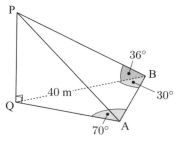

Find:

a the height of the pole, PQ

b the distance between A and B.

26 In the diagram opposite, AD = DC = BD = 4 cm, and angle BDC = 64°.

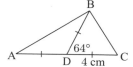

a Show that $A\widehat{B}C$ is a right angle.

b If BC has length 4.24 cm, find the length of AB.

c Find the area of △ABC.

27 Consider the quadrilateral ABCD shown.

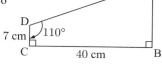

a Show that AB has length 21.56 cm correct to two decimal places.

b Find the length AD.

c Find the area of quadrilateral ABCD.

28 The area of triangle DEF is 120 cm².

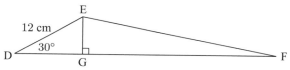

a Show that DF is 40 cm.

b Find the length of the perpendicular from E to DF.

29 Monument X is observed from two points B and C which are 330 m apart. $X\widehat{B}C$ is 63° and $B\widehat{C}X$ is 75°.

a Draw a neat, labelled diagram to illustrate this information.

b Find the distance to the monument from B.

30 The 5th hole at the Flagstaff golf course has the layout shown. From T to A on the fairway, the distance is 240 m, and from A to F the distance is 135 m.

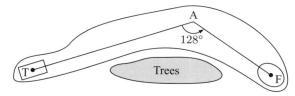

a Find the distance from T to F in a straight line.

b Find angle ATF.

31 Triangle ABC has AB = 8 cm, BC = 10 cm, and AC = 12 cm.

a Draw a neat, labelled diagram showing this information.

b Find the smallest angle in triangle ABC.

c Find the area of triangle ABC.

32 Quadrilateral PQRS has the measurements shown.

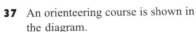

a Find the length of PR.

b Determine the measure of the angle marked θ.

33 A rectangular prism has height 7 cm, and its length is twice its width. Its volume is 350 cm³.

a Show that the prism has dimensions 5 cm by 10 cm by 7 cm.

b Find the length of the longest pencil that will fit inside the prism.

34 Triangle ABC has AB = 17 cm, AC = 19 cm, and its area is 120 cm².

a Draw a diagram to display this information.

b Determine the measure of BÂC.

c Determine the length of the remaining side.

d Find the volume of a triangular prism with cross section triangle ABC, and with length 13.5 cm.

35 A square-based right pyramid has base side length 50 m and vertical height 7 m.

a Draw a diagram of the pyramid.

b Calculate the length of a slanting edge.

c Find the measure of the angle between the slanting edge and the diagonal across the base.

d Find the surface area of a triangular face of the pyramid.

36 Quadrilateral ABCD has
AB̂C = AD̂C = 90°.
AD = 10 cm, DC = 6 cm, and BC = 9 cm.

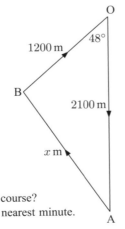

a Find the length of AB.

b Determine the area of quadrilateral ABCD.

c If this quadrilateral was the cross section of a solid prism with volume 484 cm³, what would the length of the prism be?

37 An orienteering course is shown in the diagram.
Competitors begin and end at O, travelling in a closed path to A and then B.

a Given that AÔB = 48°, calculate the distance x. Give your answer correct to the nearest metre.

b Calculate OÂB.

c If Jerry jogs at an average speed of 8 km h⁻¹, how long will it take him to complete the course? Give your answer correct to the nearest minute.

38 Lin wants to calculate the height of a mobile phone tower. He measures the angle of elevation from point A to the top of the tower C, then moves 60 m closer to point B and takes a second measurement. The information is given in the diagram below.

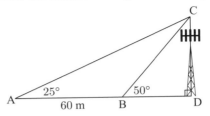

a Calculate the measure of angle ACB.

b Determine the height of the mobile phone tower.

39 An equilateral triangle ABC has sides of length 50 cm. Perpendicular lines are drawn from B to AC and from A to BC. These two lines meet at point D. DM is perpendicular to AB.

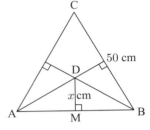

a Write down the measure of AB̂C.

b Calculate the shortest distance, x cm, from AB to D.

40 A 6 metre flagpole is supported by three guy wires. Each wire is fixed to the ground 3.2 metres from the base of the flagpole, and attached $\frac{2}{3}$ of the way up the pole.

a Draw a diagram to show this information.

b Find the angle of elevation of each wire.

c Find the total length of all three wires.

SKILL BUILDER - LONG QUESTIONS

1 The vertices of triangle ABC are A(7, 1), B(−3, −5), and C(−1, 3).

a Plot the points A, B, and C on a set of coordinate axes. Use a scale of 1 cm to represent 1 unit on each axis.

b Find the length of AC.

c Show that the length of BC is equal to the length of AC.

d The gradient of BC is 4. Find the gradient of the line through A and C.

e Show that AĈB = 90°.

f Find the area of triangle ABC.

g Determine the equation of the line through A and B. Express your answer in the form $ax + by = d$ where $a, b, d \in \mathbb{Z}$.

h D is the midpoint of AB. Find the equation of the line through D and C.

2 A regular dodecagon (12-sided polygon) is inscribed in a circle of radius 6 cm. Points A and B are adjacent vertices of the dodecagon, and both lie on the circle.

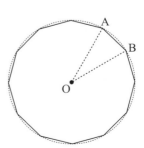

a Deduce that AÔB = 30°.

b Show that the area of triangle AOB = 9 cm².

c Hence, determine the area of the dodecagon.

d Write down the exact area of the circle, giving your answer in the form $k\pi$ where $k \in \mathbb{N}$.

e Find the resulting percentage error if:

   i the area of the circle is approximated by the area of the dodecagon

   ii the circumference of the circle is approximated by the perimeter of the dodecagon.

3 **a** Plot the points A(-5, -2), B(3, 4), and C(5, -4) on a set of coordinate axes.

Use a scale of 1 cm to represent 1 unit for both axes.

b Find M, the midpoint of AB.

c Find N, the midpoint of BC.

d Find the gradient of the straight line through M and N.

e Hence find the equation of the straight line through M and N. Express your answer in the form $ax + by = d$ where $a, b, d \in \mathbb{Z}$.

f Show that AC is parallel to MN.

g MN is $\sqrt{26}$ units long. Show that AC is twice this length.

h AB is 10 units long and BC is $\sqrt{68}$ units long.

   i Find the measure of angle ABC.

   ii Show that the area of triangle ABC is 38 units^2.

   iii The area of triangle BMN is 9.5 units^2. Find the perpendicular distance between the parallel sides of quadrilateral AMNC.

4 ABCD is a parallelogram in which A is $(0, 2)$, D is $(-3, -2)$, and C is $(4, 0)$.

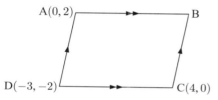

a Write down the coordinates of M, the midpoint of AC.

b Given that the midpoint of AC coincides with the midpoint of BD, determine the coordinates of B.

c Calculate the length of the line segment AD.

d Calculate angle $A\widehat{D}C$.

e Hence, or otherwise, determine the area of parallelogram ABCD.

5 A manufacturer produces wooden door-stops with the shape of the triangular prism shown.

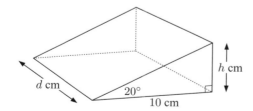

a Calculate the height h correct to 4 significant figures.

b Determine the area of the triangular cross-section.

c Given that the volume of the door-stop is 60 cm^3, determine its depth d.

d Calculate the total surface area of one triangular prism. Give your answer correct to 3 significant figures.

e A resin costing \$1.20 per m^2 is used to coat the surface of the door-stop. Determine the cost of coating 10 000 door-stops with resin. Give your answer correct to the nearest 10 dollars.

6 The vertices of triangle ABC are A(-4, 4), B(10, 2), and C(2, -4).

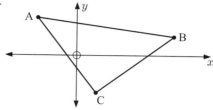

a Find the lengths of AC and BC.

b Show that triangle ABC is isosceles.

c Find the coordinates of the midpoint M of the line AB.

d State the measure of angle AMC.

e The gradient of the line through A and C is $-\frac{4}{3}$.

   i Find the gradient of the line BC, and hence show that angle ACB $= 90°$.

   ii Find the size of $C\widehat{B}A$.

f Find the equation of the line joining B and C. Express your answer in the form $ax + by = d$ where $a, b, d \in \mathbb{Z}$.

g Write down the equation of a line that is:

   i parallel to the line through B and C

   ii perpendicular to the line through B and C.

7 The shape of Clifton Park forms quadrilateral ABCD as shown below. A diagonal path crosses the park from A to C.

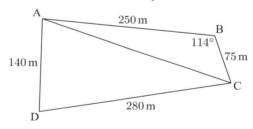

a **i** Calculate the length of the diagonal AC.

   ii Find the measure of angle ADC.

   iii Find the area of triangle ADC.

   iv Determine the area of the whole park.

b Inside the park are two identical towers in the shape of right circular cones. Each tower has vertical height 7.3 m and base diameter 3 m.

   i Find the slant height of each structure.

   ii Find the angle between the base and the slant.

   iii Find the total volume of the two conical towers.

   iv The curved surface of each cone is to be painted. One litre of paint will cover 15 m^2. Calculate the number of litres of paint required to paint both conical towers.

8 **a** Three footballers form a triangle and kick a ball between them. Juan is 35 metres from Lee and 40 metres from Kim. Kim is 50 metres from Lee.

   i Draw a neat, labelled diagram to show this information.

ii Through what angle must Kim turn after receiving the ball from Juan, to kick it to Lee?

b An archaeologist has located some ancient treasure at point T in a desert. To mark the location he uses his theodolite to measure the direction of the treasure from two points 60 metres apart. His measurements are shown below.

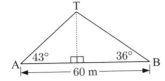

   i Find angle ATB.

   ii Find the distance from A to T.

   iii Find the shortest distance from the line AB to the treasure T.

   iv The triangle marked by A, T, and B is considered a sensitive area. How large is the sensitive area?

9 The points A, B, and C have coordinates $(-2, 1)$, $(2, 2)$, and $(3, 6)$ respectively. Point M is the midpoint of AC.

  a Plot the triangle ABC.   **b** Calculate the length of AC.

  c Find the coordinates of M.

  d Calculate the measure of $\widehat{MBC}$.

  e Find the gradient of AC.   **f** Find the gradient of BM.

  g Determine the equation of the line BM.

10 A farmer owns a triangular field ABC. Side AC has length 104 m, side AB has length 65 m, and the angle between these two sides is $60°$.

D is the point on BC such that AD bisects the $60°$ angle. The farmer divides the field into two parts A_1 and A_2 by constructing a straight fence AD of length x m.

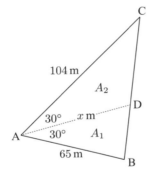

  a Use the cosine rule to calculate the length of BC.

  b Find the total area of the field.

  c **i** Find the area of A_1 in terms of x.

     ii Find a similar expression for the area of A_2.

     iii Hence, find the value of x.

  d Given that the length of BD is $\frac{5}{8}$ of the length of DC, calculate the size of $\widehat{ADB}$.

11 The diagram below shows the routes of two triangular orienteering courses ABC and DEF. The distance from E to F is 20% longer than the distance from A to C.

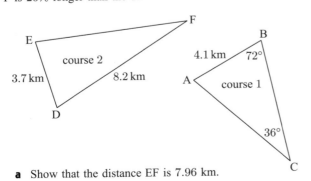

  a Show that the distance EF is 7.96 km.

b Find the measure of angle DEF.

c Find the total area covered by course 2.

d Find the length of course 1.

e Wael is able to run at an average of 14 km h^{-1} on course 1. Course 2 has more hills, and he is only able to average 10 km h^{-1}. Calculate the additional time it will take him to complete course 2 compared with course 1. Give your answer to the nearest 5 minutes.

12 A large artificial ice cream for a shop front display is to be made with a hemisphere on top of an inverted cone.

The total height of the structure is 7 m, and the cone is 4 m in height.

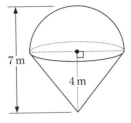

  a Show that the radius of the cone is 3 m.

  b Calculate the total volume of the ice cream.

  c Find the angle between the cone and the base of the hemisphere.

  d Find the slant height of the cone.

  e Find the total surface area of the ice cream.

  f The ice cream is to be made from a lightweight polymer, with each square metre of material weighing 1.23 kg. Calculate the total weight of the ice cream.

13 The diagram alongside shows a circle with centre O and radius 12 cm. The chord AB subtends an angle of $75°$ at the centre. The tangents to the circle at A and at B meet at P.

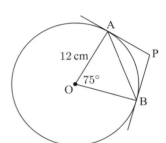

  a Using the cosine rule, find the length of AB.

  b Find the area of triangle OAB.

  c Given $\widehat{OAP} = \widehat{OBP} = 90°$, calculate the length of BP.

  d Calculate the area of triangle ABP.

14 A street sign will be made as a regular hexagon with equal sides of length 20 cm.

From the centre of the sign, line segments are drawn to the vertices. Each line bisects each vertex.

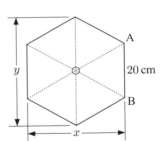

  a Show that $\widehat{AOB} = 60°$.

  b What is the nature of triangle OAB?

  c Find the total area of the sign.

  d The sign will be made from a rectangular piece of metal, and the remainder will be discarded.

    i Calculate the overall height of the sign, y.

    ii Calculate the overall width of the sign, x.

    iii Find the area of the rectangular piece of metal the sign will be cut from.

    iv What proportion of the metal is wasted?

    v Each square metre of metal costs €350. Calculate the cost of the metal that is wasted from the cutting process.

A **relation** is any set of points on the Cartesian plane.

A **function** is a relation in which no two different ordered points have the same x-coordinate or first member.

We use a **vertical line test** to see if a relation is a function.

Consider a function $y = f(x)$.

- the **domain** of f is the set of permissible values that x may take
- the **range** of f is the set of permissible values that y may take
- the x-intercepts are the values of x which make $f(x) = 0$
- the y-intercept is $f(0)$.

LINEAR FUNCTIONS

A **linear function** has the form $f(x) = mx + c$ where m and c are constants.

The graph of $y = f(x)$ is a straight line with gradient m and y-intercept c.

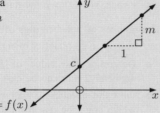

QUADRATIC FUNCTIONS

A **quadratic function** has the form $f(x) = ax^2 + bx + c$ where a, b, and c are constants, $a \neq 0$.

The graph of $y = f(x)$ is a **parabola**.

The graph has a vertical **axis of symmetry** $x = -\dfrac{b}{2a}$ and **vertex** or **turning point** at $\left(-\dfrac{b}{2a}, f\left(-\dfrac{b}{2a}\right)\right)$.

If $a > 0$, the graph opens upwards and the vertex is a local minimum.

If $a < 0$, the graph opens downwards and the vertex is a local maximum.

If the graph has x-intercepts, the axis of symmetry lies midway between them.

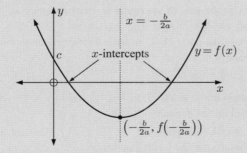

You should be able to use technology to find points at which:

- a linear function meets a quadratic
- two quadratic functions meet.

EXPONENTIAL FUNCTIONS

In this course you need to deal with two forms of exponential function:

- $f(x) = ka^x + c$ which is increasing for all x
- $f(x) = ka^{-x} + c$ which is decreasing for all x.

In each case:

- a and k control the steepness of the curve
- $y = c$ is the equation of the **horizontal asymptote**.

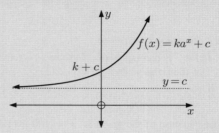

Exponential functions are commonly used to model **growth** and **decay** problems.

Exponential equations are equations where the variable appears in an index or exponent.

You should be able to solve exponential equations using technology.

PROPERTIES OF FUNCTIONS

You should be able to use a graphics calculator to:

- graph a function
- find axes intercepts
- find asymptotes
- find domain and range
- find turning points
- find where functions meet.

An **asymptote** is a line which a function gets closer and closer to but never meets.

We observe *horizontal* asymptotes in the exponential functions.

We observe *vertical* asymptotes in **power functions** such as $f(x) = \dfrac{1}{x}$ and $f(x) = \dfrac{1}{x^2}$.

SKILL BUILDER - SHORT QUESTIONS

1 The graph below shows the cost $\$C$ of a telephone call which lasts t minutes.

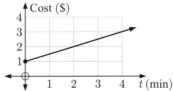

  a Find the equation of the line $C(t)$.

  b Using your equation, calculate the cost for a call lasting 23 minutes.

  c Determine the length of a phone call which costs $\$18.31$. Round your answer to the nearest minute.

2 Consider the function $h(t) = 1 - 2t^2$.

  a Find $h(0)$.

  b Find the values of t for which $h(t) = 0$.

  c Write down the domain of h.

  d Write down the range of h.

3 The graph below shows the percentage P of radioactive Carbon-14 remaining in an organism t thousands of years after it dies.

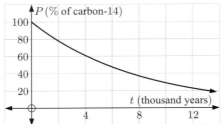

a Use the graph to estimate:

   i the percentage of Carbon-14 remaining after 4 thousand years

   ii the number of years for the percentage of Carbon-14 to fall to 50%.

b The equation of the graph is $P = 100 \times (1.1318)^{-t}$ for all $t \geqslant 0$.

   Calculate the percentage remaining after 19 thousand years.

c Write down the equation of the asymptote to the curve.

4 Consider the functions $f(x) = 15 - 2x$ and $g(x) = 2^x + 1$.

  a Calculate $f(2)$.   **b** Calculate $g(-2)$.

  c Find the value of x for which $g(x) = f(x)$.

5 The data below shows the price P that each bag of rice can be purchased for at a wholesale market in Jakarta if b bags are bought.

b bags	30	35	40	45	50
P rupiah	38 000	36 000	34 000	32 000	30 000

  a Determine the function $P(b)$.

  b *Hence* find the total cost of purchasing 60 bags of rice.

6 The graph of a quadratic function has x-intercepts 2 and -1, and passes through the point $(3, 12)$.

  a Find the equation of the function in the form $y = ax^2 + bx + c$.

  b Using an algebraic method, find the vertex of the graph of the function.

7 The exponential function $y = a \times 2^x + b$ passes through the following points:

x	0	1	2	3
y	20	p	35	q

  a Write down two linear equations which could be used to determine the values of a and b.

  b Solve the linear equations simultaneously to find a and b.

  c Hence, find the values of p and q.

8 Consider the function $k(t) = 2t - 4$ for $0 \leqslant t < 4$, $t \in \mathbb{Z}$.

  a List the elements of the domain of $k(t)$.

  b List the elements of the range of $k(t)$.

  c Sketch the function k on a set of axes, showing all elements in the domain and range.

9 The sum of the first n terms of an arithmetic sequence can be found using the formula $S_n = \dfrac{n}{2}\left(2a + (n-1) \times d\right)$, where a is the first term of the sequence and d is the common difference.

  a The first term of an arithmetic sequence is 7 and its common difference is -5. If the sum of the first n terms of the sequence is -1001, show that n can be found by solving the equation $5n^2 - 19n - 2002 = 0$.

  b Solve the equation in **a** to find n.

10 The number of people N on a small island t years after settlement, increases according to the formula $N = 120 \times (1.04)^t$. Use this formula to calculate:

  a the number of people who started the settlement

  b the number of people present after 4 years

  c the number of years it will take for the number of people to double from the initial settlement.

11 A function f is defined as $f(x) = \sqrt{x + 4}$ for $-4 \leqslant x \leqslant 12$, $x \in \mathbb{R}$.

  a Calculate: **i** $f(-4)$ **ii** $f(0)$ **iii** $f(12)$.

  b Sketch $y = f(x)$.

  c Hence, write down the range of $f(x)$.

12 The number of ball bearings of diameter d mm which can be cast from a given quantity of metal is given by $N = kd^{-3}$.

  a Determine the value of k if 810 ball bearings with diameter 2 mm can be cast.

  b Find the number of ball bearings with diameter 3 mm which could be cast from this quantity of metal.

  c If 23 ball bearings of radius r could be cast, find the value of r.

13 The population growth of a hive of bees is given by $P(t) = 120(2.25)^{\frac{t}{3}}$, where t represents time in weeks.

  a Sketch $P(t)$ for $0 \leqslant t \leqslant 20$.

  b Find the population of bees in the hive after 10 weeks.

  c Find how long it takes for the hive to number more than 5000 bees.

14 A graph of the quadratic $y = ax^2 + bx + c$ is shown alongside, including the vertex V and y-intercept.

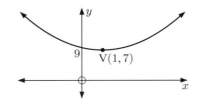

  a Determine the value of c.

  b Use the axis of symmetry to write an equation involving a and b.

  c Use the point $(1, 7)$ to write another equation involving a and b.

  d Find a and b.

15 Consider the function $f(x) = x^3 - 3x^2 - x + 3$, where f is defined on the domain $-2 \leqslant x \leqslant 3$, $x \in \mathbb{R}$.

  a Sketch the graph of $y = f(x)$, showing any axes intercepts and turning points.

  b Determine the range of f.

16 An object thrown vertically has height (relative to the ground) given by $H(t) = 19.6t - 4.9t^2$ metres after time t seconds.

  a Determine $H(0)$ and interpret its meaning.

b Find the time when the object returns to ground level.

c Write down the domain of $H(t)$ using interval notation.

d Determine the maximum height reached by the object.

17 A part of the graph of the function $f(x) = x^2 + 3x - 28$ is shown below.

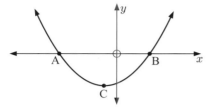

a Expand and simplify the expression $(x+7)(x-4)$.

b *Hence* find the coordinates of A and B.

c Determine the equation of the axis of symmetry.

d Write down the coordinates of C, the vertex of the parabola.

18 Consider the function $f(x) = 8x - 2x^2$.

a Factorise fully $f(x)$.

b Determine the x-intercepts for the graph of $y = f(x)$.

c Write down the equation of the axis of symmetry.

d Determine the coordinates of the vertex.

19 The function $g(x) = 4 \times 2^x - 3$ is defined for all $x \in \mathbb{R}$. Write down the range of g.

20 The number of ants in a colony is given by $N(t) = 30 - 27 \times 3^{-t}$ thousand, where t is the time in months after the beginning of the colony.

a Calculate the initial number of ants in the colony.

b Calculate the population of ants after 2 months.

c Find the time taken for the colony to reach 20 000 ants.

d Determine the equation of the horizontal asymptote of $N(t)$.

e According to the function $N(t)$, what is the smallest number of ants the colony will *never* reach?

21 Consider the function $f(x) = \dfrac{2^x}{x-1}$.

a Find the y-intercept.

b Determine the minimum value of $f(x)$ for $x \geqslant 1$.

c Write down the equation of the vertical asymptote.

d Calculate $f(5)$.

e Sketch the graph of $y = f(x)$ for $-4 \leqslant x \leqslant 7$ on a set of axes like those given, showing all the features found above.

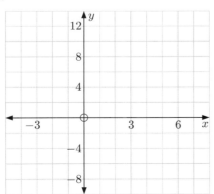

22 The diagram shows the graph of the quadratic function $f(x) = x^2 + mx + n$, including the vertex V.

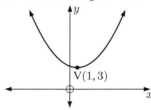

a Determine the values of m and n.

b Find k given that the graph passes through the point $(3, k)$.

c Find the domain and range of $f(x)$.

23 Consider the function $y = 3 + \dfrac{1}{x-2}$.

a Sketch a graph of the function on a grid like the one below.

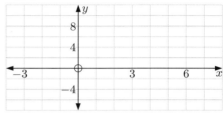

b Write down the equations of the vertical and horizontal asymptotes.

24 Before it is turned on, a refrigerator has an internal temperature of 27°C. Three hours later it has cooled to 6°C.

The internal temperature T (in °C) is given by the function $T(t) = A \times B^{-t} + 3$, where A and B are constants and t is measured in hours.

a Given that $T(0) = 27$, determine the value of A.

b Hence determine the value of B.

c Find the internal temperature of the refrigerator 5 hours after being turned on.

d Write down the minimum temperature that the refrigerator could be expected to reach.

25 The diagrams below are sketches of four out of the following five functions. Match each diagram with the correct function.

A $y = a^x$ **B** $y = a - x$ **C** $y = \dfrac{1}{x}$

D $y = x^2 - a$ **E** $y = x - a$

a   **b**

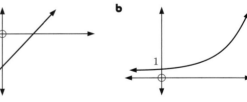

c   **d**

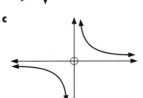

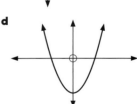

26 A quadratic function has the form $f(x) = ax^2 + bx + 7$. It is known that $f(2) = 7$ and $f(4) = 23$.

a Construct a set of simultaneous equations involving a and b.

b Find a and b.   **c** Hence calculate $f(-1)$.

27 The daily profit made by a business depends on the number of workers employed. It can be modelled by the function $P(x) = -50x^2 + 1000x - 2000$ euros, where x is the number of workers employed on any given day.

   a Using a set of axes like those below, sketch the graph of $P(x) = -50x^2 + 1000x - 2000$.

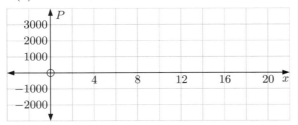

   b Determine the number of workers required to maximise the profit.

   c Find the maximum possible profit.

28 Identify the graphs which best represent the functions $f(x)$, $g(x)$, and $h(x)$.

   a $f(x) = x^2 - 3x$ **b** $g(x) = x^2 + 3x$

   c $h(x) = x^2 - 9$

A
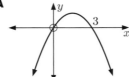
B
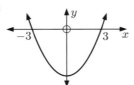

C
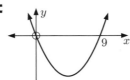
D
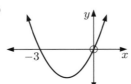

E
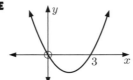

29 Consider the graph of $y = f(x)$, where $f(x) = x^3(x - 2)(x - 3)$.

   a Use your graphics calculator to find:

      i all x-intercepts

      ii the coordinates of any turning points.

   b Sketch the graph of $y = f(x)$ using a grid like the one given. Show all of the features found above.

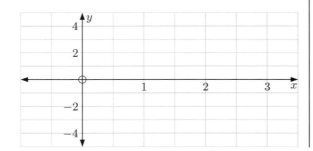

30 A graph of the form $y = a(x - 1)(x - 5)^2$ is shown below.

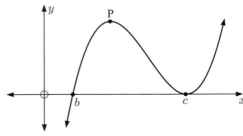

   a Determine the values of b and c.

   b Given that the y-intercept is -50, calculate the value of a.

   c Find the coordinates of P, which is a local maximum.

31 **a** Sketch the function $y = x^4 - 2x^3 - 3x^2 + 8x - 4$ for $-3 \leqslant x \leqslant 3$ using an appropriate scale.

   b Write down the coordinates of any turning points, and add these points to your graph in **a**.

   c Write down the value(s) of x for which $x^4 - 2x^3 - 3x^2 + 8x - 4 = 0$.

32 Let $f(x) = 3 - 4^{-x}$.

   a Points A$(2, p)$ and B$(-2, q)$ lie on $y = f(x)$. Determine p and q.

   b For the graph of $y = f(x)$, determine the:

      i x and y intercepts

      ii equation of the horizontal asymptote.

   c Sketch the graph of $y = f(x)$, showing all detail from above.

   d Write down the range of $f(x)$.

33 Consider $h(x) = x^2 - 2^{-x} + \dfrac{1}{x}$.

   a Determine $h(-2)$.

   b Solve $h(x) = 2$.

   c Write down the equation of the vertical asymptote.

   d Sketch $y = h(x)$, illustrating your results from **a**, **b**, and **c**.

   e Determine the range of $y = h(x)$.

SKILL BUILDER - LONG QUESTIONS

1 A company manufactures and sells DVD players.

If x DVD players are made and sold each week, the weekly cost to the company is $C(x) = x^2 + 400$ dollars.

The weekly income obtained by selling x DVD players is $I(x) = 50x$ dollars.

   a Find:

      i the weekly cost for producing 20 DVD players

      ii the weekly income when 20 DVD players are sold

      iii the profit or loss incurred when 20 DVD players are made and sold.

   b Find an expression for the weekly profit $P(x) = I(x) - C(x)$ when x DVD players are produced and sold. Do not simplify or factorise your answer.

   c The maximum weekly profit occurs at the vertex of the function $P(x)$.

      Determine the number of DVD players which must be made and sold each week to gain the maximum profit.

d Calculate the profit made on *each* DVD player if the company does maximise the profit.

e Given that $P(x)$ can be written as $P(x) = (x - 10)(40 - x)$, find the largest number of DVD players the company could produce each week and still make a positive profit.

2 Consider the graph of $y = f(x)$ where $f(x) = 2 + \dfrac{4}{x + 1}$.

  a Find the x-intercept.   **b** Find the y-intercept.

  c Calculate $f(-2)$.

  d Determine the equation of the:

    i horizontal asymptote   **ii** vertical asymptote.

  e Sketch the graph of $y = 2 + \dfrac{4}{x + 1}$ for $-5 \leqslant x \leqslant 3$.
    Label the axes intercepts and asymptotes clearly.

3 A rectangular field with one side x m is enclosed by a 160 m long fence.

  a Write down the width of the field in terms of x.

  b Write a function $A(x)$ for the area of the field.

  c Sketch the graph of $y = A(x)$.

  d Find the dimensions of the field which has the largest possible area for the given length of fence.

  e The actual area of the field is 1200 m².

    i Find the dimensions of the field.

    ii The average production yield for this field is 6.5 kg m^{-2}. Determine the amount of production lost by not using the best possible dimensions for the fence.

4 A population of insects grows according to the rule $N = 1200 \times k^t$, where N is the number of insects and t is the time in months since the population was first measured.

  a Given that $N = 4800$ when $t = 4$, find the value of k.

  b Complete the following table of values, giving your answers correct to the nearest 10 insects.

t	0	1	2	3	4	5	6
N					4800		

  c Draw the graph of N for $0 \leqslant t \leqslant 6$. Use 2 cm to represent 1 month on the horizontal axis, and 1 cm to represent 1000 insects on the vertical axis.

  d Use your graph to find the number of insects present after $2\frac{1}{2}$ months. Give your answer correct to the nearest 100 insects.

  e Find the number of months it takes for the population to reach 20 000 insects.

  f Calculate the percentage change in the number of insects in the sixth month.

5 The graph of $y = \dfrac{4}{2x - 3}$ is shown below.

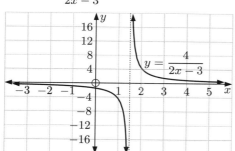

a Write down the equation of the:

  i vertical asymptote   **ii** horizontal asymptote.

b   **i** The graph of $y = a^x$ where $a > 0$, is to be drawn on the axes above. Determine the y-intercept for this graph.

    ii Determine the equation of the horizontal asymptote for $y = a^x$.

c Determine the value of a such that $y = a^x$ intersects $y = \dfrac{4}{2x - 3}$ when $x = 2$.

d For the value of a calculated in **c**, sketch the graph of $y = a^x$ on a set of axes like those given above.

6 Consider the function $f(x) = \dfrac{5}{x} - 1$ for $x \in \mathbb{R}$.

  a For what value of x is $f(x)$ undefined?

  b Find the asymptotes of $y = f(x)$.

  c The points $A(1, m)$ and $B(n, 0)$ lie on the graph of $y = f(x)$. Calculate the values of m and n.

  d Sketch the graph of $y = f(x)$ for $0 \leqslant x \leqslant 8$. Clearly show all features found in parts **a**, **b**, and **c**.

  e   **i** Show that $\dfrac{5}{x} - 1 = 5 - x$ can be rearranged into the form $(x - 1)(x - 5) = 0$.

    ii Hence solve the equation.

  f The linear function $h(x) = cx + d$ passes through the points A and B.

    i Write down the values of c and d.

    ii Add the graph of $y = h(x)$ to your sketch in **d**.

  g Write down the *positive* values of x for which $h(x) \geqslant f(x)$.

  h A quadratic function $g(x)$ passes through the origin and the points A and B. Determine the values of p, q, and r given that $g(x) = px^2 + qx + r$.

7 The monthly cost $\$C$ for a mobile phone depends on the total outgoing call time t minutes for the month. Jane compares three different monthly mobile phone plans:

  Plan 1: Fixed fee of $50

  Plan 2: Fixed fee of $20 plus $0.60 per minute of outgoing calls.

  Plan 3: No fixed fee but $2.00 per minute of outgoing calls.

  a Use a graph to illustrate the costs of the three monthly plans. Your axes should show $0 \leqslant t \leqslant 60$ and $0 \leqslant C \leqslant 120$. Let 1 cm represent 5 minutes on the t-axis, and 1 cm represent $10 on the C-axis. Clearly label each plan.

  b Calculate the monthly cost of each plan if Jane were to use her mobile phone for outgoing calls totalling:

    i 30 minutes   **ii** 60 minutes.

  c In terms of C and t, write down the equation of:

    i *Plan 1*   **ii** *Plan 2*   **iii** *Plan 3*

  d Determine the coordinates of all points of intersection between the three graphs. Clearly mark these coordinates on your graph in **a**.

  e Complete the following statements:

    i *"Plan 1 is the least expensive provided ...... minutes of outgoing calls are made."*

    ii *"Plan 2 is the least expensive provided ...... minutes of outgoing calls are made."*

iii *"Plan 3 is the least expensive provided ...... minutes of outgoing calls are made."*

8 The cost of producing x bicycles in a factory is given by $C(x) = 6000 + 40x$ pounds. The revenue earned from the sale of these bicycles is given by $R(x) = 100x$ pounds.

   a Graph each function on a set of axes like those below.

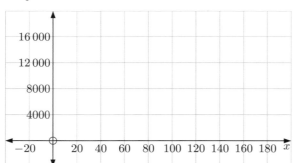

   b Determine the initial setup cost before any bicycles are produced.

   c Calculate the number of bicycles which must be produced and sold in order for the factory to break even.

   d Calculate the revenue earned *per bicycle*.

   e Write down the profit function $P(x)$ for all $x \geqslant 0$.

   f Find the profit resulting from the sale of 400 bicycles.

9 Consider the function $f(x) = \dfrac{x^3}{2^x}$.

   a Use your graphics calculator to find:

     i the y-intercept

     ii the maximum value of $f(x)$ for $-2 \leqslant x \leqslant 12$

     iii $f(2)$ and $f(-1)$.

   b Sketch the graph of $y = f(x)$ for $-2 \leqslant x \leqslant 12$ on a set of axes like those shown below, showing all of the features found above.

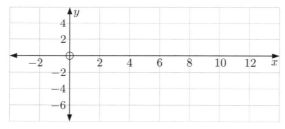

   c On the same axes as above, sketch the function $g(x) = (x - 5)^2 - 7$.

   Clearly show the coordinates of the minimum for $g(x)$, and any intersection points with $f(x)$.

   d Hence, or otherwise, find all solutions to $\dfrac{x^3}{2^x} + 7 = (x - 5)^2$.

10 The length l, width w, and height h of a rectangular prism are all measured in centimetres. The volume of the prism is fixed and equal to 2000 cm^3, and the length is double the width.

   a Show that $h = \dfrac{1000}{w^2}$.

   b Hence, show that the surface area A of the prism is given by $A(w) = 4w^2 + \dfrac{6000}{w} \text{ cm}^2$.

   c Write down the domain of $A(w)$.

   d Draw the graph of $A(w)$ for $0 \leqslant w \leqslant 30$ and $0 \leqslant A \leqslant 5000$. Clearly show the coordinates of any turning points.

   e Using your graph, or otherwise, write down the:

     i minimum possible surface area of the prism

     ii dimensions which minimise the surface area.

   f Write down the range of $A(w)$.

11 Consider the four graphs **A**, **B**, **C**, and **D** below.

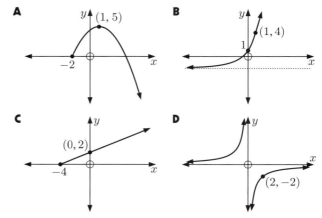

   a Which graphs, if any, have the same domains? Explain your answer.

   b Write down the range of each graph.

   c Copy and complete the table below, matching each graph above to one of the equations listed below.

Graph	Equation
	$y = a(x - h)^2 + k$
	$y = mx + c$
	$y = p \times q^x + r$
	$y = \dfrac{v}{x}$

   d Using the information shown on each graph, and the equations from **c**, determine the value of each unknown constant a, h, k, m, c, p, q, r, and v.

TOPIC 7:   CALCULUS

RATES OF CHANGE

A **rate** is a comparison between two quantities of different kinds.

The **average rate of change** between two points on a graph is the **gradient of the chord** between them.

The **instantaneous rate of change** at a particular instant is the **gradient of the tangent** to the graph at that point.

APPROXIMATING THE GRADIENT OF A TANGENT TO $y = f(x)$

For the graph of a function $y = f(x)$, consider two points $A(x, f(x))$ and $B(x + h, f(x + h))$.

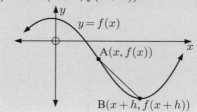

The gradient between A and B is $\dfrac{f(x+h)-f(x)}{h}$.

This gradient can be used to approximate the gradient of the tangent to $y=f(x)$ at the point A. As the point B is brought closer to the point A, the approximation improves.

THE DERIVATIVE FUNCTION

Given a function $f(x)$, the **derivative function** $f'(x)$ gives the gradient of the tangent to the curve for any value of x.

If we are given y in terms of x, the derivative function is $\dfrac{dy}{dx}$.

Function	$f(x)$	$f'(x)$
a constant	a	0
x^n	x^n	nx^{n-1}
a constant multiple of x^n	ax^n	anx^{n-1}
multiple terms	$u(x)+v(x)$	$u'(x)+v'(x)$

TANGENTS TO CURVES

Consider the tangent to $y=f(x)$ at the point where $x=a$.

To find the equation of the tangent:

- find the coordinates of the point of contact
- find the gradient function $f'(x)$ and evaluate $f'(a)$
- determine the equation of the line passing through $(a, f(a))$ with gradient $f'(a)$, using:

$$y - f(a) = f'(a)(x-a)$$

NORMALS TO CURVES

A **normal** to a curve is a line which is perpendicular to the tangent at the point of contact.

The **gradient of the normal** at $x=a$ is $-\dfrac{1}{f'(a)}$.

INCREASING AND DECREASING FUNCTIONS

Suppose a function $f(x)$ is defined on an interval S, and that we can find the derivative $f'(x)$.

For the functions we deal with in this course, $f(x)$ is:

- **increasing** on S if $f'(x) \geqslant 0$ for all $x \in S$
- **strictly increasing** on S if $f'(x) > 0$ for all $x \in S$
- **decreasing** on S if $f'(x) \leqslant 0$ for all $x \in S$
- **strictly decreasing** on S if $f'(x) < 0$ for all $x \in S$.

You should be able to use a **sign diagram** to help identify intervals where a function is increasing or decreasing.

STATIONARY POINTS

A **stationary point** of a function is a point such that $f'(x) = 0$.

It could be a **local minimum**, **local maximum**, or **horizontal inflection**.

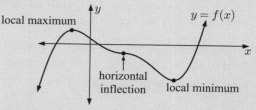

You should understand the difference between local minima and maxima, and global minima and maxima.

You should be able to use sign diagrams to identify stationary points.

Stationary point	Sign diagram of $f'(x)$ near $x=a$	Shape of curve near $x=a$
local maximum	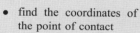	
local minimum		
horizontal or stationary inflection		

APPLICATIONS OF CALCULUS

You should be able to apply calculus techniques to find **rates of change**, including speed, cost, revenue, and profit functions.

$\dfrac{dy}{dx}$ gives the rate of change in y with respect to x.

If y increases as x increases, then $\dfrac{dy}{dx}$ will be positive.

If y decreases as x increases, then $\dfrac{dy}{dx}$ will be negative.

You should be able to use calculus to **optimise** a function on a given domain.

Step 1: Draw a large, clear diagram of the situation.

Step 2: Construct a formula with the variable to be optimised as the subject. It should be written in terms of **one** convenient variable, x say. You should write down what restrictions there are on x.

Step 3: Find the **first derivative** and find the values of x when it is **zero**.

Step 4: If there is a restricted domain such as $a \leqslant x \leqslant b$, the maximum or minimum may occur either when the derivative is zero, or else at an endpoint.

Show using the **sign diagram test** or the **graphical test**, that you have a maximum or a minimum situation.

SKILL BUILDER - SHORT QUESTIONS

1 Consider the function $y = x^3 - 4.5x^2 - 6x + 13$.

   a Find $\dfrac{dy}{dx}$.

   b Calculate the x-coordinates of the points on the function where the tangent has a gradient of 6.

2 Consider the graph of $y = ax^2 + bx + c$ where a, b, $c \in \mathbb{R}$ and $a \neq 0$.

   a Find $\dfrac{dy}{dx}$ in terms of a, b, and c.

   b When $x = k$, $\dfrac{dy}{dx} = 0$. Find the value of k in terms of a, b, and c.

   c It is known that for $x < k$, $\dfrac{dy}{dx} < 0$, and that for $x > k$, $\dfrac{dy}{dx} > 0$.

     i Is y increasing or decreasing when $x < k$?

     ii Does $x = k$ correspond to a local maximum or a local minimum?

3 **a** Write $y = \dfrac{7}{x^3}$ in the form $y = 7x^k$, where $k \in \mathbb{Z}$.

   b Differentiate $y = \dfrac{7}{x^3}$, giving your answer in the form $y = \dfrac{a}{x^b}$ where $b \in \mathbb{N}$.

4 Consider the graph of $y = \dfrac{2x^4 - 4x^2 - 3}{x}$.

   a Write the function as three separate terms in the form ax^k, $k \in \mathbb{Z}$.

   b Differentiate y with respect to x.

   c Find the gradient of the function at $x = -1$.

5 Consider the function $f(x) = 3x^2 - \dfrac{4}{x} + 7$.

   a Find $f'(x)$.   **b** Hence, find $f'(2)$.

6 The graph of $y = f(x)$ is shown. The tangent to $f(x)$ is horizontal when $x = 2$.

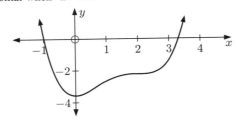

   a State intervals for which $y = f(x)$ is:

     i increasing   **ii** decreasing.

   b Without finding the function $f(x)$, find x such that $f'(x) = 0$.

   c On a diagram like the one given, draw the tangent to $y = f(x)$ at $x = 3$.

7 Consider the function $f(x) = x^4 - 6x^2 - x + 3$.

   a Find $f'(x)$.

   b Find the x-coordinates of any maximum or minimum values of $f(x)$.

8 Consider $g(x) = \left(3x - \dfrac{1}{x}\right)^2$.

   a Fully expand and simplify $g(x)$.

   b Hence, find $g'(x)$.

   c Find the gradient of $y = g(x)$ at $x = -1$.

9 **a** Differentiate the function $y = 2x^3 - 3x^2 - 264x + 13$.

   b Hence, find the x-coordinates of the points where the gradient of $y = 2x^3 - 3x^2 - 264x + 13$ is equal to -12.

10 Find the equation of the normal to $y = x^2 + 1$ at the point where $x = 2$.

11 Consider the function $f(x) = 3x^3 + \dfrac{4}{x} - 3$.

   a Find $f'(x)$.

   b Find the gradient of the tangent to $f(x)$ at $x = -2$.

   c Determine the equation of the tangent to $f(x)$ at the point where $x = -2$.

12 **a** Differentiate $y = x^2 - 6x + 9$.

   b The gradient of the tangent to $y = x^2 - 6x + 9$ at $x = a$ is -2.

     i Find the value of a.

     ii Find the point on the curve where $x = a$.

13 **a** Find the equation of the tangent to $y = 3 - x + x^2 - 2x^3$ at the point where $x = -1$.

   b Use your graphics calculator to find where this tangent cuts the curve again.

14 The curve $y = x^3 + ax + b$ has a horizontal tangent at the point $(2, -5)$.

   a Find $\dfrac{dy}{dx}$ in terms of a and b.

   b Find the values of a and b.

15 Consider the curve $y = f(x)$ where $f(x) = \dfrac{x+1}{x^2}$, $x < 0$.

   a Find $f'(x)$.

   b Use your calculator to find a local minimum point.

   c Sketch the graph of $y = f(x)$ for $-5 \leqslant x < 0$.

16 **a** Expand the expression $(2x - 1)(x^2 - 1)$.

   b Hence differentiate $f(x) = (2x - 1)(x^2 - 1)$ with respect to x.

   c Find the equation of the normal to $f(x)$ at the point where $x = -1$.

17 Consider the function $f(x) = \dfrac{3}{x^2} + x - 4$.

   a Differentiate $f(x)$.

   b Find $f'(1)$ and explain what it represents.

   c Determine the x-coordinate of the point where the gradient of the curve is zero.

18 Consider the function $f(x) = ax^2 + bx + c$, where a, b, $c \in \mathbb{Z}$.

   a Find an expression for $f'(x)$.

   b Given that $f'(x) = 2x - 6$, find the values of a and b.

   c The minimum value of f is 2. Determine the value of x which gives this minimum value.

   d Hence, find the value of c.

19 The velocity of an object after t hours is modelled by $v(t) = 7t - t^2$ km h^{-1} for $0 \leqslant t \leqslant 7$.

    a Find the velocity when $t = 1$.

    b At what other time on $0 \leqslant t \leqslant 7$ is the velocity the same as when $t = 1$?

    c Find $v'(t)$.

    d Hence, find the maximum velocity of the object.

20 A tangent to the curve $y = x^2 - kx + 11$ where $k \in \mathbb{Z}$, is known to pass through A$(6, 7)$ and B$(0, -5)$.

    a Find the gradient of this tangent.

    b Find $\dfrac{dy}{dx}$ in terms of k.

    c Let C be the point of contact where the tangent meets the curve. If C has x-coordinate 4, find the value of k.

21 Consider the function $f(x) = x^2$.

    a Find $f(2)$.   **b** Find $f(2 + h)$.

    c Find an expression for $m = \dfrac{f(2 + h) - f(2)}{h}$ in simplest form.

22   **a** Differentiate $y = \dfrac{4}{x} + x$ with respect to x.

    b Calculate the values of x for which $\dfrac{dy}{dx} = 0$.

    c Hence, find the coordinates of any local maximum or minimum values.

23 The height H of a small toy aeroplane, t seconds after it is thrown from the top of a building, is given by the function $H(t) = 80 - 5t^2$ metres, where $t \geqslant 0$.

    a Find the initial height of the toy aeroplane.

    b Determine the time it takes for the toy aeroplane to hit the ground.

    c Find $H'(t)$.

    d Find $H'(2)$ and explain what this value means.

24   **a** Write the expression $\dfrac{4 - 3x + x^2}{x^2}$ as the sum of three fractions.

    b Find $\dfrac{d}{dx}\left(\dfrac{4 - 3x + x^2}{x^2}\right)$.

    c Solve $\dfrac{d}{dx}\left(\dfrac{4 - 3x + x^2}{x^2}\right) = 0$.

25 Consider the curve $y = \dfrac{a}{x} - x^2 + 1$ where $a \in \mathbb{R}$.

    The gradient of the curve is -5 when $x = 2$.

    a Find the value of a.

    b Hence, find the point on the curve where $x = 2$.

    c Determine the equation of the tangent to the curve at $x = 2$.

26 The distance s of a group of hikers from their base camp t hours after starting out is given by $s = 2t^3 - 18t^2 + 54t$ km, for $0 \leqslant t \leqslant 5$.

    a Calculate the average velocity of the hikers in the first 4 hours.

    b Find $\dfrac{ds}{dt}$.

    c Find the values of t for which $\dfrac{ds}{dt} = 0$.

    d Hence determine the distance travelled by the hikers before they stop momentarily within the first 4 hours.

27 Consider the function $f(x) = x^3 - 3x^2 - 24x + 26$.

    a Find $f'(x)$.

    b Find the values of x such that $f'(x) = 0$.

    c Find the gradient of the normal to $y = f(x)$ at the point where $x = 1$.

28 Consider the function $f(x) = \frac{2}{3}x^3 - 2x^2 - 5x + 10$.

    a Find $f'(x)$.

    b Evaluate $f'(0)$.

    c Write down the equation of the tangent to $y = f(x)$ at $x = 0$.

    d Using technology to assist you, determine the coordinates where the tangent at $x = 0$ intersects the curve again.

29 Consider the function $f(x) = 2x^2 - 4x - 5$.

    a Find $f'(x)$.

    b Copy and complete the following table:

x	-3		6
$f'(x)$		0	20

    c The point M lies on $y = f(x)$ at the point where $f'(x) = 0$.

      i Write down the coordinates of M.

      ii Describe the point M in relation to the curve $y = f(x)$.

30 Find the coordinates of the points on the curve $y = 3x^3 + 1$ where the gradient of the normal is $-\frac{1}{9}$.

31 The graph of $f(x) = x + \dfrac{1}{x}$ is given for $x > 0$.

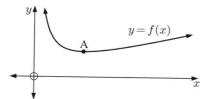

    a Find $f'(x)$ and solve the equation $f'(x) = 0$.

    b Find the coordinates of the local minimum A.

    c Copy and complete:

    "the sum of a positive number and its reciprocal is at least ......"

    d How many positive solutions would these equations have?

      i $x + \dfrac{1}{x} = 1$ **ii** $x + \dfrac{1}{x} = 2$ **iii** $x + \dfrac{1}{x} = 3$

    Give reasons for your answers.

32 A graph of $y = f(x)$ is shown below. Points A, B, C, and D lie on the curve.

a Copy and complete the table below by filling in either *positive*, *negative*, or *zero*.

	A	B	C	D
$f(x)$				
$f'(x)$				

b Sketch $y = f'(x)$.

33 A and B are two points on the curve $y = f(x)$ where $f(x) = 4\sqrt{x+3}$.

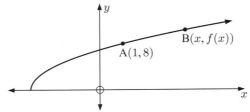

Suppose $M(x)$ generates the gradient of the chord between A and B.

a Write down the domain of $f(x)$.

b Determine an expression for $M(x)$.

c Using your expression for M, calculate the values of p and q in the table alongside. Round your values to 4 decimal places.

d Suggest the value of the gradient of $f(x)$ at point A.

x	$M(x)$
2	p
1.5	0.9706
1.1	0.9938
1.01	q
1.001	0.9999

34 Find the equation of the normal to $y = 2x^2 - \dfrac{3}{x}$ at the point where $x = -0.5$.

35 A cubic polynomial has the form $y = x^3 - 3x^2 - 24x + 7$.

a Determine $\dfrac{dy}{dx}$.

b Hence, determine all values of x which satisfy $\dfrac{dy}{dx} = 21$.

c Given that $\dfrac{dy}{dx} = 0$ for $x = -2$ and $x = 4$, determine all intervals where $f(x)$ is decreasing.

36 Consider the function $f(x) = 4x^2 + \dfrac{1}{x}$.

a Write down $f'(x)$.

b Using an algebraic method, show that $f(x)$ has only one local maximum or minimum, and that this occurs at $x = \frac{1}{2}$.

c A sketch of $y = f(x)$ is shown below.

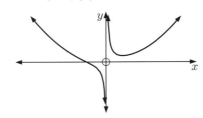

On what intervals is $f(x)$ decreasing?

37 The graph below shows the function $y = k^x$, where $k > 0$ and $x \in \mathbb{R}$.

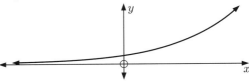

a On what interval is $y = k^x$ increasing?

b Find the y-intercept of the graph.

c The gradient of $y = k^x$ at $x = 0$ is $m = 1$. Find the equation of the normal at $x = 0$.

38 a Using calculus, determine the gradient of the tangent to the curve $y = x^3 - x + 2$ at the point where $x = -1$.

b The tangent to the curve at $x = k$ is parallel to the tangent at $x = -1$. Find k.

39 Consider the function $f(x) = \dfrac{6}{x} - \dfrac{2}{x^3}$.

a Find $f'(x)$.

b Find the values of x for which $f'(x) = 0$.

c Find the equations of the tangents to $y = f(x)$ at the points where $f'(x) = 0$.

40 The distance travelled by a truck after t hours is given by $s(t) = t^4 - 12t^3 + 48t^2 + 4t$ km.

a Find $s'(t)$.

b Evaluate $s'(3)$ and interpret its meaning.

c Find $\dfrac{s(3) - s(0)}{3 - 0}$.

d Interpret your answer to **c**.

41 Consider the function $f(x) = \frac{1}{3}x^3 - x^2 - 15x + 15\frac{2}{3}$.

a Differentiate $f(x)$ with respect to x.

b Calculate the values of x where $f'(x) = 0$.

c With the help of technology, sketch the graph of $y = f(x)$. Indicate clearly the local maximum, local minimum, and y-intercept.

42 Consider the function $y = x^3 - 3x + 2$.

a Find $\dfrac{dy}{dx}$.

b Hence, find the position of the local maximum and local minimum points.

c Sketch the graph of $y = x^3 - 3x + 2$, clearly showing the maximum and minimum points, and the y-intercept.

SKILL BUILDER - LONG QUESTIONS

1 Consider the function $f(x) = 4x^2 - 3x - \dfrac{2}{x}$.

a Calculate $f(-1)$.

b Differentiate $f(x)$ with respect to x.

c Calculate the gradient of $y = f(x)$ at $x = -1$.

d Hence, determine the equation of the tangent to $y = f(x)$ at $x = -1$.

e Use your graphics calculator to find the coordinates of the other point where this tangent meets the curve.

2 The cost of producing x bracelets is modelled by the function $C(x) = -0.2x^2 + 4x + 10$ dollars for $0 \leqslant x \leqslant 10$.

a Calculate $C(5)$ and explain what this represents.

b Differentiate C with respect to x.

c State the units that $C'(x)$ would be measured in.

d Find $C'(5)$.

e Is C increasing or decreasing at $x = 5$?

f Determine the maximum value of C on $0 \leqslant x \leqslant 10$.

3 Consider the cubic function $y = x^3 + ax^2 + bx + 3$, where a and b are integers.

a Given that the point $(1, 8)$ lies on $y = x^3 + ax^2 + bx + 3$, find an equation connecting a and b.

b Determine $\dfrac{dy}{dx}$ in terms of a and b.

c The tangent at $(1, 8)$ has equation $y = 2x + 6$.

  i Determine the gradient of the tangent at $x = 1$.

  ii Write down a second equation connecting a and b.

d Hence, determine the values of a and b.

4 The volume of water in a tank while filling is given by $V(t) = 10t^2 - \frac{1}{3}t^3$ litres, where t is the time in minutes and $0 \leqslant t \leqslant 30$.

a Find $V(5)$ and explain what this represents.

b Find $V'(t)$ and express your answer in fully factorised form. Do not forget to include units.

c Find t when $V'(t) = 0$.

d Find $V'(5)$ and $V'(25)$.

e Determine the range of values for which $V(t)$ is increasing.

5 The function $f(x) = ax + \dfrac{b}{x}$ has the tangent $y = 3x + 2$ at $x = 1$.

a Find the coordinates of the point at which the tangent touches $f(x)$.

b Hence, write an equation involving a and b.

c Find $f'(x)$ in terms of a and b.

d Calculate the gradient at $x = 1$.

e Using the results from **c** and **d**, write a second equation involving a and b.

f Find the values of a and b.

6 A closed cylinder has height h cm and base radius r cm.
The surface area of the cylinder is $10\,000$ cm^2.

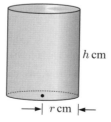

a Show that $h = \dfrac{5000 - \pi r^2}{\pi r}$.

b Hence, show that the volume of the cylinder can be written as $V = 5000r - \pi r^3$ cm^3.

c Write down $\dfrac{dV}{dr}$.

d Use calculus to determine the radius which maximises the volume of the cylinder. Give your answer correct to 4 significant figures.

e Hence, calculate the maximum volume correct to the nearest 100 cm^3.

7 Consider the function $f(x) = x^3 - 12x^2 + 36x + 5$.
A graph of $y = f(x)$ is shown below, where A is a local maximum and B is a local minimum.

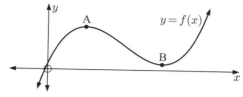

a Find $f'(x)$.

b Solve $f'(x) = 0$.

c Hence, determine the coordinates of: **i** A   **ii** B

d Line P is a tangent to $y = f(x)$ at A. Write down the equation of line P.

e Line Q is perpendicular to line P and passes through point B. Write down the equation of line Q.

f Lines P and Q intersect at point C.

  i Write down the coordinates of C.

  ii Determine the area of the triangle ABC.

  iii Calculate the size of $\widehat{\text{CAB}}$.

8 A rectangular dog enclosure is to be constructed against a wall.

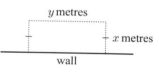

a Given that 30 metres of fence will be used, show that $y = 30 - 2x$.

b Hence, find an expression for the area of the enclosure $A(x)$ in terms of x only.

c Suppose the area of the enclosure is to be 100 m^2. Determine the dimensions of the enclosure.

d   **i** Using differential calculus, find the dimensions of the largest possible enclosure that can be created with the 30 metres of fence.

  ii Find the largest possible area of the enclosure.

9 Consider the graph of $y = f(x)$ where $f(x) = x^3$.

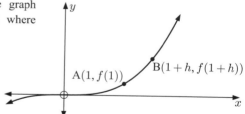

a Point A is located on $y = f(x)$ at $x = 1$. Find the coordinates of A.

b Point B is located at $x = 1 + h$. Find the coordinates of B if:

  i $h = 1$ **ii** $h = 0.1$ **iii** $h = 0.01$

c Hence, find the gradient of AB if:

  i $h = 1$ **ii** $h = 0.1$ **iii** $h = 0.01$

d Which of these values of h would provide the best approximation to the gradient of the tangent to $y = f(x)$ at $x = 1$?

e By finding the derivative of $f(x) = x^3$, find the actual gradient of the tangent to $y = f(x)$ at $x = 1$.

10 A rectangular box with an *open* top is to be constructed from cardboard as shown. There is 1200 cm² of cardboard available.

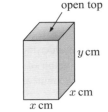
open top
y cm
x cm
x cm

a Find an expression for the total surface area of the box.

b Hence, show that $y = \dfrac{1200 - x^2}{4x}$.

c Find an expression for the volume V in terms of x only.

d Find $V'(x)$.

e Hence find the dimensions which maximise the volume of the box.

f Determine the maximum volume.

11 A farmer wants to enclose a rectangular section of his property. There is a good straight fence running down one side, so he needs to build new fences on the other three sides of the rectangle.

Suppose the farmer uses x metres of the existing good fence, as shown below.

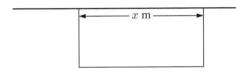
←——— x m ———→

a Suppose the farmer uses y metres of fence to complete the other 3 sides. Write an expression involving x and y to represent the length of the shorter side.

b If the enclosure has area 450 m², show that
$$y = x + \frac{900}{x}.$$

c Find $\dfrac{dy}{dx}$.

d Find the value of x which minimises y.

e Calculate the minimum amount of fencing the farmer has to buy.

f State the dimensions of the finished enclosure.

12 When a student throws a ball, its height at any time t seconds is given by $h(t) = 2.5 + 12.5t - 2.6t^2$ metres, $0 \leqslant t \leqslant 5$.

a Calculate the height of the ball at the moment it was thrown.

b Find the height of the ball after 5 seconds and explain the meaning of your answer.

c Find $\dfrac{dh}{dt}$.

d Calculate $\dfrac{dh}{dt}$ at the moment the ball was thrown.
Explain the meaning of your answer.

e Show that the ball reaches its maximum height when $t \approx 2.40$ seconds.

f The student was trying to throw the ball over a building which is 18 m high. Did the ball ever reach this high? Explain your answer.

TRIAL EXAMINATION 1

Paper 1 (1 hour 30 minutes)

1 Consider the set of numbers
$$\{\sqrt{3}, \tfrac{1}{3}, 5, \pi, -5, \sqrt{16}, 0.\overline{6}\}.$$
Write down the numbers from the set which also belong to the set of:

a integers $\mathbb{Z}$ **b** rational numbers $\mathbb{Q}$

c natural numbers $\mathbb{N}$.

2 On average, a basketballer scores 3 times out of every 5 attempts.

a Find the probability that in two attempts, the player will:
  i score both times   **ii** miss at least once.

b In the next game, the basketballer has 30 attempts at goal. How many times do you expect the player to be unsuccessful?

3 The exact measurements of the length and width of a rectangular swimming pool are 9.85 m and 5.90 m respectively.

a Calculate the area of the surface of the swimming pool.

b Jonty rounds each measurement to the nearest whole number before he approximates the area.
Determine the percentage error in Jonty's calculation.

4 30 students were asked the type of milk drinks they like. 8 students said they like plain milk only, 13 like chocolate milk only, and 4 students do not like either.

a Find the number of students who like both plain and chocolate milk.

b Represent the information above on a Venn diagram.

c Determine the probability that a student, chosen at random from this group, likes only one type of milk.

5 The frequency table below shows the price of tickets for various events at the Festival of Arts.

Cost ($)	Number of events
20 - 39	12
40 - 59	15
60 - 79	11
80 - 99	7
100 - 119	5

a Find the probability that a ticket for a randomly chosen event at the festival costs $60 or more.

b Find the mean and standard deviation for the price of tickets.

c Find the percentage of events whose ticket price is less than 0.722 standard deviations above the mean.

6 Andy is travelling from New York to London. He can buy 1.00 USD for 0.68 GBP, and sell 1.00 USD for 0.64 GBP.

a If Andy converts 2000 USD into GBP, how many GBP will he receive?

b Andy spends 1100 GBP while in London, then converts the remainder back to USD. How many USD will he receive?

7 The seventh term of a geometric sequence is 320, and the tenth term is 2560. Find:

a the common ratio   **b** the first term

c the twentieth term.

8 Ali purchased her car at the start of 2010 for €12 000. She borrowed all of the money from her bank at a nominal rate of 8.5% p.a., compounding monthly. The amount is to be repaid over 3 years.

  a Calculate the total amount that Ali will repay.

  b If the average annual rate of depreciation for the car is 22.5%, calculate its value at the end of 2013.

9

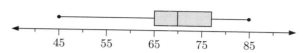

Consider the box plot above.

  a Find the upper and lower quartiles.

  b Calculate the: **i** interquartile range   **ii** range.

  c What proportion of values are less than 70?

10 A company sells plastic boxes for \$12.50 each. It estimates that the total cost of producing x boxes is $(9.5x + 45)$.

  a Find the profit made when 100 boxes are produced and sold.

  b Find the number of boxes which must be produced and sold for the firm to 'break even'.

  c Find the number of boxes which must be produced and sold to make a profit of \$1000.

11 The truth table below shows some truth values for the statement $(p \veebar q) \Rightarrow \neg(p \wedge q) \vee q$.

$p \wedge q$	$\neg(p \wedge q)$	$p \veebar q$	$\neg(p \wedge q) \vee q$	$(p \veebar q)$ $\Rightarrow \neg(p \wedge q) \vee q$
T	F	F		
F	T	T	T	T
F	T	T	T	T
F	T			T

  a Fill in the missing truth values on the table.

  b Write down the contrapositive of the proposition:
  If Bozo is a clown then Bozo has a red nose.

12 **a** Find the equation of the line joining the points $A(-2, -3)$ and $B(1, 3)$. Give your answer in the form $ax + by + d = 0$ where $a, b, d \in \mathbb{Z}$.

  b Find the equation of the perpendicular bisector of AB.

13 A function is defined by $f(x) = ax^2 + bx + d$ where a, b, and d are integers.

  a Write an expression for $f'(x)$.

  b If $f'(x) = 5x - 10$, find the values of a and b.

  c The minimum value of $f(x)$ is -4.
  Determine the x-coordinate of the minimum value of $f(x)$, and hence find the value of d.

14 A megaphone in the shape of a cone has vertical angle $60°$ and a slant height of 45 cm, as shown in the diagram.

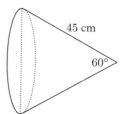

45 cm

60°

  a Determine the diameter of the megaphone.

  b Determine the volume of the megaphone.

15 In a recent examination, the scores were normally distributed with mean 52 and a standard deviation of 12.

  a The top 8% of candidates were awarded an A, and the next 25% were awarded a B. Find the minimum score required to receive a B, giving your answer to the nearest integer.

  b If 300 students sat for the examination, find the expected number who scored less than 40 marks.

Paper 2 (1 hour 30 minutes)

1 The coordinates of the vertices of a parallelogram are $A(-2, 1)$, $B(6, 3)$, $C(3, -1)$, and $D(-5, d)$. AC is a diagonal of the parallelogram.

  a Using a scale of 1 cm to represent 1 unit on both axes, plot the points A, B, and C.   *(2 marks)*

  b **i** Find the gradient of the line through B and C.

    ii Explain why the gradient of AD is the same as the gradient of BC.

    iii Find the value of d.   *(5 marks)*

  c The length of AC is $\sqrt{29}$ units, and the length of BC is 5 units.

    i Find the length of AB.

    ii Find the measure of angle ABC.   *(6 marks)*

  d Find the area of parallelogram ABCD.   *(3 marks)*

2 A sports locker at school contains 6 basketballs, 9 footballs, and 5 volleyballs. During a sports lesson, a teacher chooses two balls from the locker at random, without replacement.

  a Draw a tree diagram showing all the possible outcomes. Write the probabilities for each branch on the diagram.   *(4 marks)*

  b Find the probability that the teacher chooses:

    i two basketballs

    ii one basketball and one football

    iii both balls the same.   *(8 marks)*

  c Suppose that instead, the teacher had chosen the balls *with* replacement. Find the probability that the balls chosen were:

    i two volleyballs

    ii both basketballs, given that the two balls were the same.   *(6 marks)*

3 The temperature of a cup of coffee in a plastic cup is modelled by $T_P(t) = 61 \times (0.95)^t + 18$ °C, where t is the time in minutes after it is poured.

  a **i** Calculate the values of a and b in the table below.

Time t (min)	0	5	10	15	20	25	30
Temperature (°C)	a	65.2	54.5	46.3	39.9	34.9	b

    ii Find the time it will take for the temperature to reach $25°$C.   *(4 marks)*

  b On graph paper, draw and label a graph to display this information. Use 1 cm for every two minutes on the horizontal axis, and 1 cm for every $10°$C on the vertical axis.   *(4 marks)*

  c A china cup is used for a new cup of coffee. The temperature of coffee in this cup is given by $T_F(t) = 53 \times (0.98)^t + 18$ °C.

i Determine the initial temperature of the new cup of coffee.

ii Comment on the rate of heat loss of the coffee in the china cup compared to the original plastic cup.

iii Using technology, find the time it takes before the temperatures in the cups are the same.
Give your answer correct to the nearest tenth of a minute. *(6 marks)*

d In the longer term, what temperature will each cup of coffee approach? *(2 marks)*

4 The following table shows the number of people using a public swimming pool over a 10 day holiday period and the forecast maximum temperature for each of these days.

Number of people	28	31	35	33	22	18	24	29	30	32
Forecast temp. (°C)	88	103	147	110	82	44	66	75	80	95

a Using technology, write down the value of the coefficient of correlation, r. *(2 marks)*

b Describe the nature of the relationship between the forecast maximum temperature and the number of people attending the swimming pool. *(2 marks)*

c **i** Write down the equation of linear regression for attendance against temperature.

ii Use this equation to estimate the number of people who will attend the pool on a day when the forecast maximum temperature is 20°C and when it is 40°C.

iii Which of your estimates above is more reliable? Explain your answer. *(8 marks)*

d The manager of the pool plans to use the forecast maximum temperature to determine the number of staff to be employed each day. State whether the manager's plan seems sensible. Justify your answer. *(2 marks)*

5 The manager of the pool records the gender of the swimmers attending the pool each day. She believes that the maximum forecast temperature on any day causes a different attendance pattern for males and females.
Given the following data about average daily attendance and temperature, conduct a chi-squared test at the 5% significance level to determine if the manager is correct.

	Temp < 30°C	Temp ⩾ 30°C
Male	36	47
Female	51	40

a Write down suitable null and alternate hypotheses for the chi-squared test. *(2 marks)*

b Show that there is one degree of freedom for this test. *(1 mark)*

c Write down the p-value of the chi-squared statistic. *(2 marks)*

d What conclusion can be drawn from this test? Justify your answer. *(2 marks)*

6 **a** Consider the function $f(x) = 3x^3 - 4x + 5$.

i Find $f(1)$. *(2 marks)*

ii Calculate $f'(x)$. *(2 marks)*

iii Find the gradient of the tangent at $x = 1$. *(2 marks)*

iv Determine algebraically the equation of the tangent to the curve at the point where $x = 1$. *(2 marks)*

v The tangent to the curve at $x = 1$ intersects the curve at one other point. Use technology to find the coordinates of this point of intersection. *(2 marks)*

b A rectangular box has the dimensions shown.

i Write an expression for the volume V of the box, in terms of x and y. *(2 marks)*

ii Given that $y = \dfrac{30\,000 - x^2}{2x}$, write the volume in terms of x only. *(2 marks)*

iii Find $\dfrac{dV}{dx}$. *(2 marks)*

iv Hence, find the value of x which maximises the volume of the box. *(3 marks)*

TRIAL EXAMINATION 2

Paper 1 *(1 hour 30 minutes)*

1 The number 0.051 762 is rounded to 0.0518.

a State which *two* of the following are accurate descriptions of the rounding:

A correct to 3 decimal places

B correct to 4 significant figures

C correct to 3 significant figures

D correct to the nearest ten-thousandth.

b Write 0.0518 in the form $a \times 10^k$ where $0 \leqslant a < 10$, $k \in \mathbb{Z}$.

c Calculate the percentage error in the rounding.

2 The histogram shows the weight of sheep in a herd.

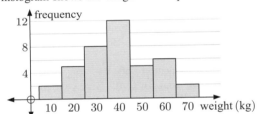

a Find the number of sheep that were weighed.

b Estimate the mean weight for these sheep.

c The farmer sends to the market all sheep whose weights are more than 50% above the mean.
Determine the percentage of the herd that will be sent to market.

3 Triangle ABC is shown in the figure with AC = 10 cm and AB = 8.5 cm. Angle CDA = 90° and angle ACD = 38°.

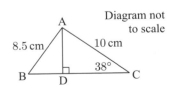

Calculate the length of: **a** DC **b** BD.

4 The cost of producing x pairs of jeans is $C(x) = 15.6x + 245$ euros. Each pair of jeans can be sold for €42.50. If 22 pairs of jeans are bought and sold, find:

a the total revenue **b** the cost of production

c the profit **d** the profit per pair of jeans.

5 P and Q are subsets of a universal set U.

$U = \{1 \leqslant x \leqslant 13, \ x \in \mathbb{Z}\}$,

$P = \{\text{prime numbers between 1 and 13 inclusive}\}$, and

$Q = \{\text{factors of 24}\}$.

List the members of sets: **a** P **b** Q **c** $(P \cup Q)'$

6 A box contains 10 wooden shapes. There are 5 triangles, 4 rectangles, and 1 rhombus. 2 shapes are chosen at random from the box without replacement.

Calculate the probability that:

a both are triangles

b one of the chosen shapes is the rhombus

c the first chosen is a rectangle, given that the second is a rectangle.

7 The weights of green sea turtle eggs are normally distributed about a mean of 40 g with standard deviation 3.45 g.

a On average, a female turtle will lay 90 eggs. Calculate the number of eggs which we expect to weigh less than 35 g.

b Estimate the upper quartile of turtle egg weights.

8 Jacinta won a prize of $6000. She spent half the money and placed the remainder in an investment account which paid a nominal rate of 7.5% per annum, compounded quarterly.

a Calculate the amount in the account after $3\frac{1}{2}$ years.

b After $3\frac{1}{2}$ years, Jacinta withdraws half of the money in the account. From this time, how many whole years will it take for the account balance to reach $4500?

9 **a** Differentiate the function $y = 2x^3 - 3x^2 - 264x + 13$.

b Hence, find the x-coordinates of the points where the gradient of the tangent to the function is equal to -12.

10 Eddie purchases 4 reams of paper and 3 pens for a total cost of £19. Let the price of a ream of paper be £r, and the price of a pen be £p.

a Write an equation involving r and p that represents Eddie's purchase.

b If the cost of 2 pens was deducted from the cost of 3 reams of paper, the total amount would be £10.

Write a second equation involving r and p that represents this information.

c Solve the two equations you have written simultaneously.

d Hence determine the cost of purchasing 5 reams of paper and 5 pens.

11 The table below shows the different activities chosen by a class of final year students on their end of year expedition.

	Climbing	Swimming	Mountain Biking
Female	9	18	8
Male	15	16	24

A χ^2 test is conducted, at the 5% level of significance, to see if the activities chosen by the students are independent of gender.

a Write down the table of expected values.

b Write down the χ^2 calculated test statistic.

c Find the p-value for the test.

d Write a conclusion for the test, giving clear reasons for your answer.

12 Twenty numbers are in an arithmetic sequence. The sum of the numbers is 560 and the last number is 66. Find:

a the first term **b** the common difference.

13 p and q are two propositions:

p: x is a prime number. q: x is a factor of 12.

a Write down $p \Rightarrow \neg q$ in words.

b Find a value of x for which $p \Rightarrow \neg q$ is false.

c Write down, in words, the inverse of $p \Rightarrow \neg q$.

d State whether the inverse is true for all values of x, giving a clear reason for your answer.

14 Consider the function $y = \dfrac{2^x}{x - 2}$.

Using your graphics calculator where necessary:

a find the value of y when $x = 3$

b find the y-intercept

c determine the minimum value of the function for $x \geqslant 2$

d write down the equation of the vertical asymptote.

15 A trapezium ABCD has non-parallel sides AB and DC. The shorter parallel side is a cm long, and the other parallel side is $2a$ cm long. The height of the trapezium is 7 cm, and its area is 42 cm^2.

a Draw a neat, labelled diagram to illustrate this information.

b Find the lengths of the parallel sides of trapezium ABCD.

Paper 2 (1 hour 30 minutes)

1 Jaime invests in a savings scheme. She puts 500 USD into the bank on the 1st January each year from 2005 to 2015 inclusive. Interest is paid on the 31st December of each year at the rate of 2.5% per annum compounded annually.

a Calculate the amount Jaime will have in the bank on:

i 2nd January 2006 **ii** 2nd January 2007.

(7 marks)

b Write down a geometric series formula for the amount of money in Jaime's account at midnight December 31st after n years. (2 marks)

c Hence find the amount that she will have in the bank at midnight on December 31st 2015. (3 marks)

d **i** Determine the total amount that Jaime invested over all these years.

ii Calculate the total interest Jaime's investments have earned. (2 marks)

2 The following list shows the results of an examination taken by 30 students in an IB class:

40	55	45	70	60	65	45	48	80	75
78	85	45	38	54	75	32	75	58	65
60	75	78	45	68	85	88	45	55	68

a Calculate the mean mark. (2 marks)

b Write down the value of the:

i median

ii lower quartile

iii upper quartile. (3 marks)

c Draw a box and whisker diagram to represent this information. (4 marks)

d **i** Display the examination results in a frequency table using the groups 30 - 39, 40 - 49, and so on.

ii Write down the modal group. (3 marks)

e **i** Estimate the mean mark using the grouped data.
    ii Find the percentage error in your estimate. (4 marks)

3 Consider the two parallel lines $L_1: y = 5 - 2x$ and
$L_2: y = kx + c$.

  a Draw the line L_1 for $-3 \leqslant x \leqslant 6$, $-3 \leqslant y \leqslant 6$.
     Use a scale of 1 cm to represent 1 unit on both axes.
                              (4 marks)

  b **i** Given that L_2 passes through $(2, -3)$, find the values
       of k and c.
    ii On your graph, draw L_2.   (5 marks)

  c **i** Show that $(3, -1)$ lies on L_1.
    ii Through the point $(3, -1)$, draw a line perpendicular
       to both L_1 and L_2. Label this line L_3.   (4 marks)

  d **i** Find the equation of L_3.
       Express your answer in the form $ax + by + d = 0$
       where $a, b, d \in \mathbb{Z}$.
    ii Determine the point of intersection of L_3 and L_2.
                               (6 marks)

4 The diagram below shows the graphs of two functions
$y = 8 - x^2$ and $y = 2^{-x}$.

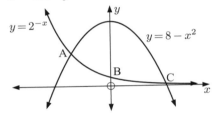

  a **i** Write down the coordinates of B, the y-intercept of
       the graph of $y = 2^{-x}$.   (2 marks)
    ii Find the coordinates of the two points of intersection,
       A and C.   (3 marks)
    iii Determine the values of x for which $2^{-x} > 8 - x^2$.
                             (2 marks)

  b Show that the equation of the tangent to $y = 8 - x^2$ at
    the point $(-1, 7)$, is $y = 2x + 9$.   (3 marks)

  c Write down the coordinates of the point where the tangent
    found in **b** meets the graph of $y = 2^{-x}$. (2 marks)

5 In a class of 40 students, 8 students like all three of the subjects
History, Science, and Music. 11 of the students like History and
Science, 10 students like History and Music, and 13 students
like Science and Music. In total 26 students like History, 21
like Science, and 16 like Music.

  a Display this information on a Venn diagram.   (4 marks)

  b Find the probability that a randomly selected student:
    i likes History only
    ii does not like Music
    iii likes Music or Science but not History
    iv likes at least one of the three subjects. (6 marks)

  c Find:
    i P(Science $\cap$ Music)
    ii P((History $\cap$ Science $\cap$ Music)$'$)
    iii P(History$'$ $\cap$ Science$'$)   (5 marks)

  d Find the probability that a student chosen at random likes
    History, given that they like Music. (2 marks)

  e Find the probability that 2 students, chosen at random, both
    like History and Science. (3 marks)

6 Consider the function $f(x) = 2x + 5 + \dfrac{3}{x}$, $x \neq 0$.

  a Write down the equation of the vertical asymptote.
                              (2 marks)

  b Find $f'(x)$.   (3 marks)

  c Find the gradient of the function at $x = \frac{1}{2}$. (2 marks)

  d Determine whether the function is decreasing or increasing
    on the interval $[\frac{1}{4}, \frac{3}{4}]$. Justify your answer. (2 marks)

TRIAL EXAMINATION 3

Paper 1   **(1 hour 30 minutes)**

1 Let $M = 6 \times 10^4$, $E = 1.5 \times 10^8$, and $I = 8 \times 10^2$.

  a Calculate $\dfrac{M}{EI}$, giving your answer in the form $a \times 10^k$
    where $1 \leqslant a < 10$, $k \in \mathbb{Z}$.

  b Find the value of $\sqrt[3]{\dfrac{34.5^2 - 103}{50.5 + 19}}$, giving your answer:
    i correct to 4 significant figures
    ii correct to the nearest whole number.

2 The temperature (in $°C$) is measured at midday in Perth over a
period of 10 days. The results are:
     18, 15, 20, 22, 18, 19, 15, 25, 24, 26
  a Calculate the mean temperature.
  b Find the median temperature.
  c A newspaper publishes the mean temperature rounded to
    the nearest whole number. Find the percentage error in
    this approximation.

3 Consider the function $y = x^3 - 6x + 2$.
  a Find the equation of the normal to the curve at the point
    where $x = 0$.
  b Write down the coordinates of the point where the normal
    in **a** meets the x-axis.

4 **a** Find the equation of the line passing through the points
    $(-2, 2)$ and $(1, 5)$.
  b The equation of a second line is $x + y = 3$. Find the
    point of intersection of the two lines.

5 The cube shown has edges
of length 20 cm.
The midpoint of AP is M.
Calculate:
  a the length of AC
  b the length of CM
  c the measure of angle CMR.

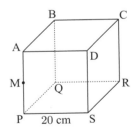

6 A series is given by $85 + 78 + 71 + \ldots - 48$.
  a Calculate the number of terms in the series.
  b Find the sum of the terms in the series.

7 A bag contains 40 beads. Some are green and the rest are
yellow. The probability of a bead drawn at random being green
is $\frac{1}{5}$.
  a Find the number of green beads in the bag.

b Calculate the probability that when two beads are drawn at random from the bag, without replacement, both will be yellow.

c How many green beads should be added to the bag to change the probability of drawing a green bead to $\frac{1}{2}$?

8 A function is given by $f(x) = x^2 + ax + b$. The zeros of $f(x)$ are $x = -1$ and $x = 3$.

a Find the values of a and b.

b Find $f'(x)$.

c Find the coordinates of the minimum point on the graph of $y = f(x)$.

9 The tables below show the number of pupils in three different groups who scored 0, 1, 2, 3, or 4 marks for a test question.

a

Marks	0	1	2	3	4
Number of pupils	5	9	x	6	8

Write down a possible value for x if the modal class is 2 marks.

b

Marks	0	1	2	3	4
Number of pupils	2	1	3	1	y

Write down a value for y if the median is 3 marks.

c

Marks	0	1	2	3	4
Number of pupils	z	4	3	2	1

Calculate z if the mean is 1 mark.

10 The points X, Z, Y, and W are on level ground. A surveyor knows that XY is 5 km, but he needs to find the distances XZ and XW.

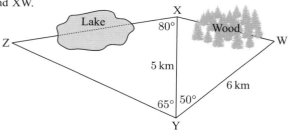

a The distance XZ cannot be measured directly because of a large lake between X and Z. By measurement he finds that angle ZYX is 65°, and angle ZXY is 80°.
Calculate XZ.

b A wood obstructs the view from X to W. The surveyor finds by measurement that YW is 6 km, and angle XYW is 50°.
Calculate XW.

11 Consider the graph of the quadratic function $g(x)$. The x-intercepts are -1 and 2. The y-intercept is 5.

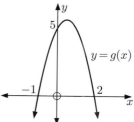

a Write down the equation of the axis of symmetry.

b Find the function $g(x)$.

12 Consider the table of values below:

x	25	37	51	41	38	32	20	36	33
y	11	20	46	30	36	24	12	24	30

a Write down the equation of linear regression for y against x.

b Hence estimate the value of y when x is 60.

c Comment on the reliability of your answer in **b**.

13 Complete the following truth tables:

a $p \vee \neg q$

p	q	$\neg q$	$p \vee \neg q$
T	T		
T	F		
F	T		
F	F		

b $(p \wedge q) \vee \neg q$

p	q	$\neg q$	$p \wedge q$	$(p \wedge q) \vee \neg q$
T	T	F		
T	F	T		
F	T	F		
F	F	T		

c State whether $p \vee \neg q$ and $(p \wedge q) \vee \neg q$ are logically equivalent. Give a clear reason for your answer.

14 Consider the function $f(x) = \dfrac{10}{x} - x^2$.

a Find $f'(x)$.

b Find the equation of the normal to the function $f(x)$ at the point where $x = -2$.

15 Water flows through a cylindrical pipe of radius 4 cm at a rate of 50 cm s^{-1}.

a Find the volume of water, in litres, that passes through the pipe in 1 hour.

b The water is used to fill a swimming pool of dimensions 10 m × 4.50 m × 1.50 m.
Calculate how long it will take to fill the swimming pool.

Paper 2 (1 hour 30 minutes)

1 a The 150 IB students at Russell High School can be members of the Performing Arts club (P), the Choir (C), or the Sports club (S). We know that:
$n(P) = 70$ $n(P \cap C) = 25$ $n(P \cap C \cap S) = 5$
$n(C) = 50$ $n(P \cap S) = 15$
$n(S) = 70$ $n(C \cap S) = 20$

i Draw a Venn diagram to illustrate this information. (4 marks)

ii Write down the number of students who are not members of any of the three clubs. (1 mark)

iii Find the probability that a student chosen at random belongs to the Performing Arts club only. (2 marks)

iv Find the probability that a student chosen at random belongs to the Performing Arts club or the Choir, but not both. (2 marks)

v It is known that a student belongs to the Choir. Calculate the probability that the student also belongs to the Sports club. (2 marks)

b At MacKenzie High School the students can only belong to one of the three clubs: Performing Arts, Choir, or Sports. The data is given in the table below.

	Performing Arts	Choir	Sports
Males	40	30	80
Females	20	50	30

A student believes that gender is a determining factor in belonging to a club. She decides to test this by conducting a χ^2 test at the 5% level of significance.

  i Write down a suitable null hypothesis.   (1 mark)

  ii Show that the expected value for females belonging to the Sports club is 44.   (2 marks)

  iii Write down the χ^2 test statistic for this data.   (2 marks)

  iv Given that $\chi^2_{crit} = 5.99$, state what the student's conclusion should be. Give a clear reason for your answer.   (2 marks)

2 The proportion of a population infected by a virus grows at a rate of 20% per day. Initially there are 10 people infected.

  a Find the number of people infected after 1 day. (2 marks)

  b Find the number of people infected after 1 week. Give your answer correct to the nearest whole number.   (3 marks)

  c The number of people infected after t days is modelled by $f(t) = Na^t$.
    State the value of: **i** N **ii** a.   (3 marks)

  d Use technology to sketch the graph of $f(t)$ for $0 \leqslant t \leqslant 20$.
    Show clearly the value of the y-intercept.   (4 marks)

  e Write down the range of $f(t)$ for the given domain. Give your answer correct to the nearest whole number.   (3 marks)

  f Write down the equation of the horizontal asymptote of the function $y = 10(1.2)^t$.   (2 marks)

3 The parallel box and whisker plot below shows the maximum daily temperatures of two cities during the month of September.

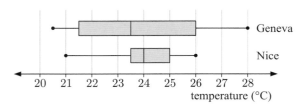

  a Write down the median temperature for Nice.   (1 mark)

  b Write down the lower quartile for Nice.   (1 mark)

  c Calculate the range and interquartile range for Nice.   (3 marks)

  d The temperature range for Geneva is $7.5°C$ and the interquartile range is $4.5°C$. Write down which city was hotter on average.   (1 mark)

  e Explain the significance of the different lengths of the two boxes.   (2 marks)

  f Find:

    i the smallest possible number of days Geneva was hotter than Nice   (3 marks)

    ii the difference between the maximum temperatures for Geneva and Nice.   (2 marks)

4 **a** The United Bank of Australia offers compound interest of 3.5% per annum on its deposit account. Niki deposits 10 000 AUD in an account at this bank.

    i Show that at the end of 5 years, the investment will total approximately 11 900 AUD.   (3 marks)

    ii Calculate the number of complete years required for Niki's 10 000 AUD to double.   (3 marks)

  b The Federal Bank of Australia offers a rate of interest which, compounded annually, will double the amount invested in only 12 years. Calculate the rate of interest offered.   (4 marks)

  c The Australian Credit Bank offers a nominal rate of interest of 3%, compounded quarterly. Sami deposits 5000 AUD. Calculate the amount of interest he will receive at the end of 6 years.   (4 marks)

  d Niki and Sami are travelling from Australia to the Netherlands. They plan to take 5000 euros with them. The Travellers Bank offers a rate of 1 euro for 1.25 AUD, and charges 1.5% commission. Calculate how many AUD they will pay for their euros.   (4 marks)

5 The graph below shows the function $f(x) = 4 - 2.5^x$.

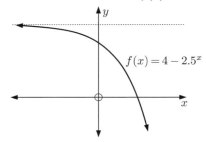

  a Write down the y-intercept.   (1 mark)

  b Find the value of x when $f(x) = 0$.   (2 marks)

  c Draw the graph of $f(x)$ for $-4 \leqslant x \leqslant 2$. Use 1 cm to represent one unit on both the x and y axes.   (4 marks)

  d Write down the equation of the horizontal asymptote.   (2 marks)

  e Write down the coordinates of the point where the graph of $f(x)$ meets the line $y = 3x - 2$.   (2 marks)

6 ABCD is a square piece of card with side length 20 cm. Four equal shaded squares are cut from the corners, each with side length x cm.

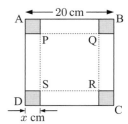

The remaining piece of card is folded along the dotted lines to form an open rectangular box with base PQRS.

  a Write down the length of PQ in terms of x.   (2 marks)

  b Show that the volume V of the box is $4x^3 - 80x^2 + 400x$.   (3 marks)

  c Find $\dfrac{dV}{dx}$.   (3 marks)

  d Find the values of x for which $\dfrac{dV}{dx} = 0$.   (3 marks)

  e State the value of x for which the volume is a maximum.   (2 marks)

Paper 1   (1 hour 30 minutes)

1 Calculate $\dfrac{\sqrt{2.068}}{1.203 \times 0.0237}$, giving the answer:

   a correct to 2 decimal places

   b correct to the nearest integer

   c correct to 6 significant figures

   d correct to the nearest ten thousandth

   e in the form $a \times 10^k$, where $1 \leqslant a < 10$ and $k \in \mathbb{Z}$.

2 Let p and q be the propositions:

     p: Joshua studied hard at Mathematics.

     q: Joshua passed Mathematics.

   a Write the following logic statements in words:

     i $\neg p \Rightarrow \neg q$ **ii** $\neg p \wedge q$

   b Write the following statement using symbols only:

     If Joshua passed Mathematics then he had studied hard.

   c Consider the compound statement in **b**. Is it the inverse, the converse, or the contrapositive of $p \Rightarrow q$?

3 **a** Before Maria left Italy for her holiday in Thailand, she converted 2400 euros to Thai baht. The exchange rate from euros to baht at the time was 1 : 37.794. The bank charges a commission of 1.5% on each transaction. Calculate the amount of baht Maria received.

   b Maria spent 90% of her baht while on holiday. When she returned home, she converted the remaining baht back to euros using the same bank. She was again charged 1.5% commission, and the exchange rate from euros to baht at that time was 1 : 36.481.

     Calculate the amount of euros she received.

4 In a class of 28 IB students, 12 study Economics, 17 study Math Studies, and 6 study neither of these.

   a Use the information above to complete the Venn diagram:

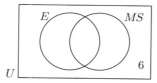

   b Find the probability that a randomly chosen student from this class studies both Economics and Math Studies.

   c A randomly chosen student from this class is known to study either Economics or Math Studies, but not both. Find the probability that this student studies Economics.

5 In triangle ABC, AB = 7.1 cm, BC = 4.7 cm, and $\widehat{ABC} = 92.7°$.

   a Calculate the length of AC.

   b Find the measure of $\widehat{CAB}$.

6 The population of a town t years after 1960 can be estimated by the formula $P = 360 \times (1.045)^t$. Estimate:

   a the population of the town in 1960

   b the population of the town in 2005

   c the year when the population reached 800.

7 The following passage contains 51 words that vary in length from 1 letter to 7 letters.

"Once a week we go to the shops. I ride in the cart while Bobby pushes. Up and down the aisles we go. We choose cereal and apples and cookies and raisins and a picture book. We pile them on the counter. Mummy pays and we all carry the bags home."

   a Complete the table below by counting the words containing 2 letters.

Number of letters	1	2	3	4	5	6	7
Number of words	3		12	11	5	5	4

   b On a grid like the one provided, construct a *histogram* to represent the data.

8 Yumiko places 2500 JPY in an investment account. Calculate the total amount of interest earned on this investment by the end of the fifth year, if the account pays:

   a compound interest of 3.2% per annum

   b 3.2% per annum nominal interest, compounded quarterly.

9 Consider the function $f(x) = ax^2 + bx + c$. The graph of the function has y-intercept 6, and passes through the points $(-2, 36)$ and $(1, -1.5)$.

   Find the values of a, b, and c.

10 The table below summarises the lengths and weights of a sample of pumpkins.

	Short	Medium	Long	Total
Light	21	15	5	41
Heavy	2	4	11	17
Total	23	19	16	58

   a One pumpkin is chosen at random. Find the probability that it is:

     i long and light

     ii not medium in length, given that it is light.

   b A second pumpkin is chosen at random without the first one being replaced. Find the probability that the two pumpkins are both short and heavy.

11 A company's profits increase by 1.5% each year for a period of 10 years. In the first year the profit was 50 000 MAR.

   a Calculate the profit in the 10th year.

   b Find the total profit for the 10 year period, giving your answer correct to the nearest 5000 MAR.

12 The length of a rectangle is 3 cm greater than its width. Let the width of the rectangle be x cm.

   a Write down the length of the rectangle in terms of x.

   b The rectangle has area 108 cm^2. Write an equation in x to represent this information.

   c Find the value of x by solving the equation.

   d Find the length and width of the rectangle.

13 David buys a new car for $45 000. At the end of the first year it has decreased in value by 20%.

    a Calculate the value of the car at the end of the first year.

    b After the first year, the car decreases further in value by 12% per year. Calculate the value of the car at the end of the third year.

    c David plans to sell the car when its value falls below one third of its cost price. Calculate the number of years before he will sell the car.

14 Consider the function $y = (x - 2)^2 - 5$.

    a Find $\dfrac{dy}{dx}$.

    b Write down the gradient of the tangent to $y = (x - 2)^2 - 5$ at the point $(3, -4)$.

    c Hence, determine the equation of this tangent.

15 A regular pentagon with sides of length 15 cm is shown below. Diagonals BD and BE are drawn to divide the figure into three triangles. Triangle BED has area 173.12 cm^2.

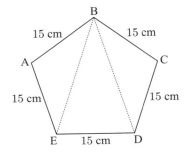

    a Determine the size of each interior angle of the pentagon.

    b Calculate the total area of the pentagon.

Paper 2   (1 hour 30 minutes)

1 At the beginning of 1990, a toy company in New Zealand began with 25 employees. Since then, the number of employees has increased by an average of 4% per year.

    a Calculate the number of employees at the beginning of:

        i 1991   **ii** 2005   (3 marks)

    b Show that there were 32 employees at the company at the start of 1996.   (1 mark)

    c The table below shows the effect of average annual rates of inflation on the value of 1 NZD since 1990.

Years since 1990	4.5%	5.5%	b%
5	1.246	1.307	1.370
10	1.553	1.708	1.877
15	a	2.232	2.572

        i Find the values of a and b.   (3 marks)

        ii The average salary paid to employees in the toy company in 1990 was 18 000 NZD. Use the table of values to determine the average salary paid in 2005 if salaries increased at 5.5% per year.   (2 marks)

        iii If the average annual salary increase was 4.5%, find the first year in which the average salary was more than 30 000 NZD.   (3 marks)

2 At the end of a golf competition, the individual scores of a number of competitors were:

68, 73, 78, 84, 71, 67, 87, 66, 73, 73, 83, 79, 67, 70, 75

    a Find the:

        i median score

        ii interquartile range of the scores.   (3 marks)

    b Represent the data using a box and whisker plot. (3 marks)

    c Calculate the mean score.   (2 marks)

    d The scores can be grouped in class intervals of 60 - 69, 70 - 79, and 80 - 89. Copy and complete the frequency table:

Score	60 - 69	70 - 79	80 - 89
Frequency	4		

        (2 marks)

    e Estimate the mean from the frequency table.   (2 marks)

    f Calculate the percentage error in the estimate of the mean from the actual mean.   (2 marks)

3 The table below shows the number of vehicles using Holden Street between 7 am and 12 noon on a weekday.

Hour	Frequency	Cumulative frequency
7:00 am to 8:00 am	14	14
8:00 am to 9:00 am	48	62
9:00 am to 10:00 am	35	
10:00 am to 11:00 am	24	
11:00 am to 12 noon	22	

    a Copy and complete the table of cumulative frequencies.   (2 marks)

    b Draw a cumulative frequency graph for this information. Use a scale of 2 units to represent 1 hour on the horizontal axis, and 1 unit to represent 20 vehicles on the vertical axis.   (4 marks)

    c Use your graph to find:

        i the time when the 'median' vehicle drove down Holden Street

        ii the time when the 100th vehicle drove down the street

        iii the number of cars which used the street between 9:30 am and 11:30 am

        iv the time for the 25th percentile.   (4 marks)

4 Two vertical poles EG and FH stand on horizontal land. The point D is 200 m to the left of the base of the pole FH. The angle from D to the top of the pole FH is 6.2°. The distance from D to the top of the pole EG is 57 m.

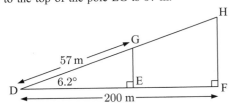

    a Find the distance EF between the bases of the poles.   (3 marks)

    b Find the slant distance GH between the tops of the poles.   (3 marks)

    c Find the size of the angle FEH, from E to the top of the pole FH.   (4 marks)

    d Determine the height of the shorter pole EG.   (2 marks)

    e Calculate the area of the triangle EGH.   (4 marks)

5 a Consider the propositions:

p: The weather is fine.
q: The bus is late.
r: I will walk to school.

i Copy and complete the last three columns of the truth table. (3 marks)

p	q	r	$(p \wedge q)$	$\neg r$	$(p \wedge q) \Rightarrow r$	$(\neg p \vee \neg q)$	$(\neg p \vee \neg q) \Rightarrow \neg r$
T	T	T	T	F			
T	T	F	T	T			
T	F	T	F	F			
T	F	F	F	T			
F	T	T	F	F			
F	T	F	F	T			
F	F	T	F	F			
F	F	F	F	T			

ii State whether the compound statements $p \wedge q \Rightarrow r$ and $\neg p \vee \neg q \Rightarrow \neg r$ are logically equivalent, tautologies, contradictions, or none of these. (1 mark)

iii Write the statement $\neg p \vee \neg q \Rightarrow \neg r$ in words. Begin with "If the weather is not fine, ...". (2 marks)

iv Write down the inverse of the statement "If the bus is late then I will not walk to school." (2 marks)

b The probability that the weather is fine on any day is 0.7. The probability that the bus is late on any day is 0.4. Find the probability of:

i the weather being fine on two consecutive days (2 marks)

ii the weather being fine and the bus not being late, on two consecutive days. (4 marks)

6 The height of an object falling towards the ground is modelled by the function $H(t) = 1600 - 4t^2$ metres, where t is the time in seconds after the object is first observed, $0 \leqslant t \leqslant 20$.

a Sketch the graph of $y = H(t)$ for $0 \leqslant t \leqslant 20$, $0 \leqslant y \leqslant 1600$. (3 marks)

b **i** Show that the object takes 20 seconds from the moment it is first observed to hit the ground. (2 marks)

ii Find the height of the object at the moment it is first observed. (2 marks)

iii Find the average speed of the object's descent to the ground, given that average speed $= \dfrac{\text{distance}}{\text{time}}$. (2 marks)

c **i** Calculate $H(10)$. (2 marks)

ii Show that $H(10 + h) = 1200 - 80h - 4h^2$. (3 marks)

iii Simplify the expression $\dfrac{H(10 + h) - H(10)}{h}$. (2 marks)

d **i** Find $H'(t)$. (2 marks)

ii State whether the function H is an increasing or a decreasing function, giving a reason for your answer. (2 marks)

iii Find the values of $H'(5)$ and $H'(10)$. (2 marks)

iv Comment on the difference in the values of $H'(5)$ and $H'(10)$. (2 marks)

Paper 1 (1 hour 30 minutes)

1 Suppose $x = 5$, $y = 12$, and $z = 100$:

a Find the value of x^2yz^3.
Write your answer in the form $a \times 10^k$ where $1 \leqslant a < 10$ and $k \in \mathbb{Z}$.

b Find the value of $\sqrt{x^2yz^3}$.
Give your answer correct to 2 significant figures.

2 In a recent test out of 40 marks, the students in a class scored the following marks:

35, 24, 28, 30, 18, 32, 38, 32, 19, 27

a Rank these marks in descending order.

b Calculate the median mark.

c Find the probability that the result for a randomly chosen student was 5 marks or less from the median mark.

d Find the probability that two students, chosen at random, both scored above 30 for the test.

3 Let $U = \{\text{all polygons}\}$,
$A = \{\text{all triangles}\}$, and
$B = \{\text{all quadrilaterals}\}$.

a Write down a subset of A.

b Write down the name of a possible element of B'.

c Represent the relationship between sets A and B on a Venn diagram.

d State the number of elements in $A \cap B$.

4 Consider two propositions p and q.

a Complete the truth table for the argument $(q \wedge \neg p) \Rightarrow (\neg q \vee p)$.

p	q	$\neg p$	$\neg q$	$(q \wedge \neg p)$	$(\neg q \vee p)$	$(q \wedge \neg p) \Rightarrow (\neg q \vee p)$
T	T	F	F			
T	F	F	T			
F	T	T	F			
F	F	T	T			

b Is the argument valid? Give a reason for your answer.

c Write down, in symbols, the contrapositive of $(q \wedge \neg p) \Rightarrow (\neg q \vee p)$.

5 a Find the x-intercepts of $y = (x - 9)(x + 3)$.

b Write down the zeros of $f(x) = 5x^2 + 53x - 84$.

c Differentiate $y = 3x^2 - 14x + 8$.

6 The time adults watch television on Saturday nights is normally distributed with mean 3.4 hours and standard deviation 42 minutes.

a In an apartment block containing 210 adults, how many do you expect to watch less than 4 hours of television on a Saturday night?

b 60% of adults watch at least k hours of television on a Saturday night. Find, to the nearest minute, the value of k.

7 a On a set of axes like the ones below, sketch the graphs of:
  i $y = \frac{3}{2}$ **ii** $2x - y - 3 = 0$

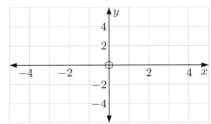

  b Find the exact value of the x-coordinate of the point of intersection of these two lines.

  c Find the distance from the line $y = \frac{3}{2}$ to the x-axis.

8 A geometric sequence is given below:
$$\tfrac{1}{64}, \tfrac{1}{32}, \tfrac{1}{16}, ..., 16, 32, 64, 128, 256$$

  a Write down the common ratio for the sequence.

  b Find the number of terms in the sequence.

  c Calculate the sum of the sequence.

9 The heights, in centimetres, of a sample of seedlings, are:

75	86	103	92	97	63	98
88	106	112	83	78	112	88
72	95	106	67	86		

  a Find the number of seedlings in the sample.

  b Write down:
   i the lower quartile   **ii** the median
   iii the upper quartile.

  c Draw a box and whisker plot for this data using a scale like the one below.

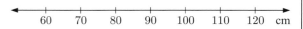

10 The cost of producing a book is given by
$C(x) = 46\,000 + 11.50x$ dollars, where x represents the number of books produced. Each book sells for $30.

  a Determine the profit if:
   i 2000 books are produced and sold
   ii 5000 books are produced and sold.

  b How many books must be sold for the publisher to break even, or in other words for the income to equal the cost?

11 Consider the scalene triangle with dimensions shown.

  a Find the measure of angle BCA.

  b Determine the area of the triangle.

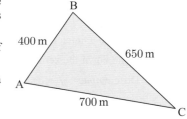

12 Philip invests 3500 pesos in an investment account.

  a Find how much the investment is worth at the end of the seventh year if the account pays:
   i 5.2% nominal interest per annum, compounding annually
   ii a nominal interest of 5.2% per annum, compounded monthly.

  b Determine the difference in the amount of interest paid in the two situations.

13 Consider the graph of the quadratic function shown. The axes intercepts are all integers.

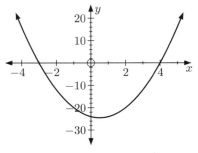

  a Write down the coordinates of the x-intercepts.

  b Determine the equation of the axis of symmetry.

  c Find the equation of the function.

  d Find the minimum value of the function.

14 Use your calculator to help answer the following questions for the data below:

P	25	45	70	55	45	90	35	45
Q	74	21	35	37	54	35	61	55

  a Write down the equation of linear regression and the coefficient of correlation.

  b Observe the scatter plot for this data. Remove the outlier and recalculate the equation of the regression line and the correlation coefficient.

  c Comment on the change in the gradient of the regression line and the strength of the relationship as a result of removing the outlier.

15 Consider the cubic function shown in the diagram.

  a Using interval notation, define where:
   i $f'(x) > 0$
   ii $f'(x) < 0$.

  b For what values of x does $f'(x) = 0$?

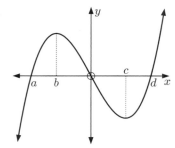

Paper 2 (1 hour 30 minutes)

1 a Alex, Carlos, and Edwin decide to move to New York. They rent a flat for the monthly rate of 700 USD each.
   i Calculate the total rent paid by the three in the first year. (2 marks)
   ii The rental agreement says that the rent will be reviewed annually and increased according to the schedule:

Year 1	Year 2
2.8%	3.3%

   Calculate the total rent increase at the end of the first year. (2 marks)
   iii Find the amount of rent each person paid in the third year. (3 marks)

  b Alex, Carlos, and Edwin work for the same company. Since the year 2000, the company has expanded its workforce by an extra 40 people each year. It is intended that the expansion will continue at the same rate indefinitely. There were 800 employees at the start of 2005.

i Find how many employees the company had at the start of 2009. (2 marks)

ii Find how many employees the company had at the start of 2000. (2 marks)

iii Calculate the year in which the company will first have more than 1000 employees. (3 marks)

2 **a** A rectangle has length 5 cm greater than its width. Let x cm be the width of the rectangle.

    i Write down the length of the rectangle in terms of x. (1 mark)

    ii Write down an expression in x for the area of the rectangle. (2 marks)

    iii The rectangle has area 204 cm². Find the value of x. (3 marks)

    iv Find the length and width of the rectangle. (1 mark)

b A hemispherical enclosure has volume 7068 m³. The enclosure is covered with waterproof material.

    i Find the radius of the enclosure. (3 marks)

    ii Calculate the amount of material covering the enclosure. (3 marks)

    iii The material costs \$23.45 for every 10 m². Find the total cost of the material covering the enclosure. (2 marks)

c A box in the shape of a cuboid has dimensions 12.7 cm by 10.8 cm by 6.45 cm. A piece of string is stretched across the inside of the cuboid.

    i Find the length of the piece of string. (4 marks)

    ii Find the angle that the piece of string makes with the base of the box in this case. (2 marks)

3 **a** There are 60 packets of assorted sweets on a stand in a shop:

42 contain peppermints

36 contain chocolates

21 contain liquorice

11 contain at least some peppermints and some liquorice

8 contain at least some liquorice and some chocolates

6 contain peppermints only

7 contain all 3 types of sweets

5 do not contain any of these 3 sweets.

    i Copy and complete the diagram using the given information.

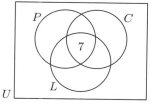

(3 marks)

    ii Find:

    (1) $n(P \cap C)$ **(2)** $n(P \cup C')$

    (3) $n((L \cap P)' \cap (C \cup L))$ (4 marks)

b **i** A box on a market stall contains assorted pieces of jewellery as shown in the table below.

	Earrings	Necklaces	Bracelets
Silver	15	8	12
Bronze	7	11	10
Gold	10	23	14

Write down the number of different pieces in the box. (1 mark)

ii Scarlett selects a piece of jewellery at random. Find the probability that the piece she chooses is:

    (1) gold earrings (2 marks)

    (2) not a silver bracelet. (2 marks)

iii Scarlett replaces the first piece and then randomly selects 2 pieces from the box. Find the probability that the pieces of jewellery she chooses are both silver given that they are both necklaces. (3 marks)

4 **a** The table below lists some values for the function $f(x) = 2x^2 - 6x - 20$.

x	-2	-1	0	1	2	3	4	5	6
y	0	a	-20	-24	b	-20	-12	0	16

    i Write down the values of a and b. (2 marks)

    ii Use the table of values above to graph $y = f(x)$ for $-3 \leqslant x \leqslant 7$. Use 1 cm to represent 1 unit on the x-axis and 1 cm to represent 10 units on the y-axis. (4 marks)

    iii On the same axes, draw the graph of $y = 2x - 1$. (2 marks)

    iv Find the coordinates of the points where $f(x)$ meets the line $y = 2x - 1$. (2 marks)

b Consider the function $f(x) = \frac{5}{3}x^3 + 3x^2 - 8x + 2$ for $0 \leqslant x \leqslant 3$.

    i Find $f'(x)$. (3 marks)

    ii Find the gradient of the normal to $y = f(x)$ at $x = 1$. (2 marks)

    iii Determine the values of x for which $f'(x) = 0$. (3 marks)

    iv Find the largest and smallest values of $f(x)$ in the given domain. (5 marks)

5 A company is testing a new fertiliser using two fields containing 100 plants each. One field has been sprayed with an old fertiliser and the other field sprayed with the new fertiliser. The results for the trials are displayed below:

Field A (old fertiliser)	
Mean plant height	23 cm
Standard deviation σ	6 cm
Number of plants $\geqslant$ mean $+ 0.368\sigma$	22
Number of plants < 25 cm	52

Field B (new fertiliser)

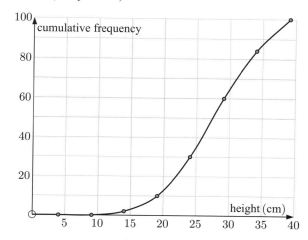

Mathematical Studies SL – Exam Preparation & Practice Guide (3rd edition)

a Using the cumulative frequency graph for field B, construct a frequency table for height h with intervals $0 \leqslant h < 5$, $5 \leqslant h < 10$, $10 \leqslant h < 15$, and so on. (2 marks)

b Use your frequency table to estimate the mean and standard deviation of the plant heights in field B. (2 marks)

c Find the percentage difference between the means for the two fields, expressed as a percentage of the mean for field B. (2 marks)

d Use the cumulative frequency graph to determine the number of plants in field B which are taller than 0.368 standard deviations above the mean. (2 marks)

6 The productivity of a group of workers was measured before and after some technical training. The results are listed below:

	Before training	After training
poor	10	16
average	48	43
good	42	41

The firm's manager believes that the training has resulted in an improvement in productivity. He conducts a χ^2 test to confirm his belief.

a State suitable null and alternate hypotheses for the χ^2 test. (2 marks)

b Calculate the χ^2 statistic for this data. (2 marks)

c Show that there are 2 degrees of freedom. (1 mark)

d The critical value for the test at the 5% level of significance is 5.991. Write a conclusion for the test, giving a reason for your answer. (2 marks)

e The manager is not happy with these results, and decides to repeat the training one week later. On this occasion, the results after training are: poor $= 6$, average $= 33$, and good $= 61$.
What conclusion can be drawn after the second training session? Justify your answer. (3 marks)

SHORT QUESTIONS

1

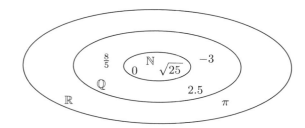

2 a

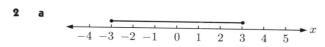

b

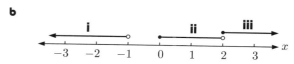

3 $\dfrac{a^2 + b}{c} = \dfrac{2.5^2 + 7}{137} \approx 0.096\,715\,328$

   a 0.10   {to 2 decimal places}

   b 0.0967   {to 3 significant figures}

   c 9.67×10^{-2} {in scientific notation}

4

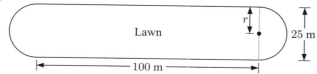

Area = area of two semi-circles + area of rectangle
$$= \pi r^2 + 100 \times 2r$$
$$= \pi \times 12.5^2 + 200 \times 12.5$$

   a Using $\pi \approx 3$, George's answer is   2968.75 m^2
$$\approx 2970 \text{ m}^2$$

   b   The actual area ≈ 2990 m^2

     $\therefore$ percentage error $\approx \left| \dfrac{2990 - 2970}{2990} \right|$

                    $\approx 0.67\%$ {to 2 significant figures}

5

	$\mathbb{N}$	$\mathbb{Z}$	$\mathbb{Q}$	$\mathbb{R}$
$(-2)^3$	N	Y	Y	Y
$\sqrt{2}$	N	N	N	Y
0.65	N	N	Y	Y
1.27×10^4	Y	Y	Y	Y

6 $\sqrt{\dfrac{32.76}{3.95 \times 2.63}} \approx 1.775\,806\,022$

   a 1.776   {to 3 decimal places}

   b 2   {to the nearest integer}

   c 1.78   {to 3 significant figures}

   d 1.78×10^0 {in standard form}

7 a Speed of light
$$\approx 186\,280 \text{ miles s}^{-1}$$
$$\approx (186\,280 \times 1.609) \text{ km s}^{-1}$$
$$\approx (186\,280 \times 1.609 \times 60) \text{ km min}^{-1}$$
$$\approx 17\,983\,471 \text{ km min}^{-1}$$
$$\approx 18\,000\,000 \text{ km min}^{-1} \text{ \{to 3 significant figures\}}$$

   b 1.80×10^7 km min^{-1}

   c The time for light to reach Earth is
$$\dfrac{\text{distance}}{\text{speed}} \approx \dfrac{195\,000\,000}{18\,000\,000} \approx 10.8 \text{ minutes.}$$

8 a Since the series is arithmetic, $a - 120 = 98 - a$

                              $\therefore 2a = 218$

                               $\therefore a = 109$

     If the common difference is d then
$$d = 109 - 120 = -11$$
$$\therefore b = 98 - 11$$
$$= 87$$

   b The sequence has $u_1 = 120$ and $d = -11$
$$\therefore u_n = 120 + (n-1)(-11)$$
$$= 131 - 11n$$

   c If $131 - 11n = 0$ then $11n = 131$
$$\therefore n \approx 11.9$$

     The last term that is positive is u_{11}, so there are 11 positive terms.

9 a For interest of 4.5% p.a. the multiplier is 1.045

     i After 1 year the value $= £18\,000 \times 1.045$
$$= £18\,810$$

     ii After 2 years the value $= £18\,810 \times 1.045$
$$= £19\,656.45$$

   b After n years the value $= £18\,000 \times 1.045^n$

   c After 13 years the value $= £18\,000 \times 1.045^{13}$
$$\approx £31\,899.53$$

   d The investment is trebled when $1.045^n = 3$
$$\therefore n \approx 24.96 \text{ years}$$

     So, the investment will have trebled after 25 years.

10 a 5.645 cm $\approx$ 5.65 cm {to 3 significant figures}

     i actual error = 5.65 cm $-$ 5.645 cm
$$= 0.005 \text{ cm}$$

     ii percentage error $= \dfrac{|\text{actual error}|}{\text{exact value}} \times 100\%$
$$= \dfrac{0.005}{5.645} \times 100\%$$
$$\approx 0.0886\%$$

   b **i** Maximum possible error $= 70 \times 0.032$
$$= 2.24 \text{ km h}^{-1}$$

     ii Minimum possible speed $= (70 - 2.24)$ km h^{-1}
$$\approx 67.8 \text{ km h}^{-1}$$

         Maximum possible speed $= (70 + 2.24)$ km h^{-1}
$$\approx 72.2 \text{ km h}^{-1}$$

11 a $-9 - (-2) = -7$ So, this is an arithmetic sequence

     $-16 - (-9) = -7$ with common difference -7.

     The sequence is $-2, -9, -16, -23, -30, ...$

b $u_n = a + (n-1)d$
$= -2 + (n-1)(-7)$
$= 5 - 7n$

12 a $1 \text{ USD} = 0.6439 \text{ GBP}$

$\therefore \quad \dfrac{1}{0.6439} \text{ USD} = \dfrac{0.6439}{0.6439} \text{ GBP}$

$\therefore \quad 1.5530 \text{ USD} \approx 1 \text{ GBP}$

The exchange rate from GBP to USD is about $1 : 1.553$

b $365 \text{ USD} = 365 \times 0.6439 \text{ GBP}$
$= 235.02 \text{ GBP}$

c $500 \text{ GBP} = 500 \times 1.5530 \text{ USD}$
$= 776.52 \text{ USD}$

13 a Commission $= 1\%$ of €1000
$=$ €10
$\therefore$ €990 $= 990 \times 125.3 \text{ JPY}$
$= 124\,047 \text{ JPY}$

So, $124\,047 \text{ JPY}$ would be received.

b Commission $= 1\%$ of $46\,000 \text{ JPY}$
$= 460 \text{ JPY}$

Amount remaining $= (46\,000 - 460) \text{ JPY}$
$= 45\,540 \text{ JPY}$

$45\,540 \text{ JPY} = 45\,540 \times \dfrac{1}{125.3} \text{ EUR}$
$= 363.45 \text{ EUR}$

So, €363.45 would be received.

14 a $u_1 = 1(1+1) = 2, \quad u_2 = 2(2+1) = 6,$
$u_3 = 3(3+1) = 12$

b $u_{15} = 15(15+1) = 240$

c If $u_n = 600$ then $n(n+1) = 600$
$\therefore \quad n^2 + n - 600 = 0$
$\therefore \quad (n-24)(n+25) = 0$
$\therefore \quad n = 24 \text{ or } -25$

But $n > 0$, so the 24th term is 600.

15

	$\sqrt{4}$	-2	3.75	π	$2.\overline{3}$
$\mathbb{N}$	✓				
$\mathbb{Z}$	✓	✓			
$\mathbb{Q}$	✓	✓	✓		✓
$\mathbb{R}$	✓	✓	✓	✓	✓

16 $2.34 + \dfrac{5.25}{3.10 \times 7.65} \approx 2.561\,378\,874$

a 3 {to the nearest integer}

b 2.5614 {to 4 decimal places}

c 2.56 {to 3 significant figures}

d 2.56×10^0 {in scientific notation}

17 a Since the series is arithmetic,
$(k - 166) - (-347) = -185 - (k - 166)$
$\therefore \quad k - 166 + 347 = -185 - k + 166$
$\therefore \quad 2k = -200$
$\therefore \quad k = -100$

b The common difference $d = (k - 166) - (-347)$
$= -100 - 166 + 347$
$= 81$
$\therefore$ since the first term $u_1 = -347$,
the general term $u_n = u_1 + (n-1)d$
$= -347 + (n-1)81$
$= -428 + 81n$

c If $-428 + 81n = 0$ then $81n = 428$
$\therefore \quad n \approx 5.28$
$\therefore$ the first positive term will be the 6th term, and this
term is $u_6 = -428 + 81 \times 6$
$= 58$

18 a $FV = PV \left(1 + \dfrac{r}{100k}\right)^{kn}$
where $r = 5.5, \ PV = $ €2500, $\ k = 4, \ n = 2$

$\therefore$ the final value $= $ €$2500 \times \left(1 + \dfrac{5.5}{100 \times 4}\right)^8$
$= $ €2788.60

b €$0.7437 = 1 \text{ USD}$

$\therefore \quad$ €$1 = \dfrac{1}{0.7437} \text{ USD}$

$\therefore \quad$ €$2788.60 = \dfrac{2788.60}{0.7437} \text{ USD}$

$= 3749.64 \text{ USD}$

19 a $FV = PV \left(1 + \dfrac{r}{100k}\right)^{kn}$
where $r = 6, \ PV = $ £14\,000, $\ k = 12, \ n = 5$
$\therefore$ the final value $= $ £$14\,000 \left(1 + \dfrac{6}{1200}\right)^{60}$
$= $ £18\,883.90

b After p periods the investment has value
$PV \left(1 + \dfrac{r}{100k}\right)^p$
$\therefore$ it is doubled when $\left(1 + \dfrac{r}{100k}\right)^p = 2$
$\therefore \quad \left(1 + \dfrac{6}{1200}\right)^p = 2$

Using technology, $p \approx 138.98$

So, the investment is doubled after 139 months, which is
11 years 7 months.

20 a The bank is selling CAD.
$\therefore \quad 300 \text{ GBP} = 300 \times 1.513 \text{ CAD}$
$= 453.90 \text{ CAD}$

b The bank is buying CAD.
$\therefore \quad 300 \text{ CAD} = 300 \times \dfrac{1}{1.589} \text{ GBP}$
$= 188.80 \text{ GBP}$

21 a $u_6 = 49 \quad$ so $\quad u_1 + 5d = 49 \quad$ .... (1)
$u_{15} = 130 \quad$ so $\quad u_1 + 14d = 130 \quad$ .... (2)
$\therefore \quad 9d = 81$
{subtracting (1) from (2)}
$\therefore$ the common difference $d = 9$

b Since $u_6 = 49, \ u_1 + 5d = 49$
$\therefore \quad u_1 + 45 = 49$
$\therefore$ the first term $u_1 = 4$

c The general term of the sequence is
$$u_n = u_1 + (n-1)d$$
$$= 4 + (n-1)9$$
$$= 9n - 5$$
If $9n - 5 = 300$ then $9n = 305$
$$\therefore \quad n \approx 33.9$$
So, the first 33 terms (only) are less than 300.

22 a Length of tubing = circumference of circle
$$= 2\pi r$$
$$= \pi d$$
$$= 4.5\pi$$
$$\approx 14.1 \text{ m}$$

b 14 m {to the nearest metre}

c Percentage error $= \left| \dfrac{14 - 4.5\pi}{4.5\pi} \right| \times 100\%$
$$\approx 0.970\%$$

23 a The common ratio $= \dfrac{6.75}{2.25} = 3$

b Since $u_1 = 0.75$, the general term $u_n = 0.75 \times 3^{n-1}$

c $S_n = \dfrac{u_1(r^n - 1)}{r - 1}$
$$\therefore \quad S_{10} = \dfrac{0.75(3^{10} - 1)}{3 - 1} = 22\,143$$

24 a For an arithmetic series, $S_n = \dfrac{n}{2}(2u_1 + (n-1)d)$
$$\therefore \quad \tfrac{7}{2}(2u_1 + 6 \times 14) = 329$$
$$\therefore \quad 7u_1 + 294 = 329$$
$$\therefore \quad 7u_1 = 35$$
$$\therefore \quad u_1 = 5$$

b If $S_n = 69\,800$ then $\dfrac{n}{2}(2u_1 + (n-1)d) = 69\,800$
$$\therefore \quad \dfrac{n}{2}(10 + (n-1)14) = 69\,800$$
Using technology, $n = 100$

25 a The exchange service is selling CHF.
400 USD = 400×0.8917 CHF = 356.68 CHF

b The exchange service is buying CHF.
356.68 CHF = $356.68 \times \dfrac{1}{0.9384} = 380.09$ USD

c The cost of the transactions is $400 - 380.09 = 19.91$ USD.

26 a $FV = PV \left(1 + \dfrac{r}{100k}\right)^{kn}$
where $r = 6.7$, $PV = £4750$, $k = 12$, $n = 4.25$
$\therefore$ the final value $= £4750 \times \left(1 + \dfrac{6.7}{1200}\right)^{12 \times 4.25}$
$$= £6309.79$$

b Interest $= £(6309.79 - 4750) = £1559.79$

c If the interest was compounded annually, $k = 1$
$\therefore$ the new interest $= £\left(4750 \times \left(1 + \dfrac{6.7}{100}\right)^{4.25} - 4750\right)$
$$= £1507.38$$
$\therefore$ the difference in interest would be
$$£(1559.79 - 1507.38) = £52.41$$

27 a time taken $= \dfrac{\text{distance travelled}}{\text{average speed}}$
$$= \dfrac{80 \text{ km}}{45 \text{ km h}^{-1}}$$
$$\approx 1.7778 \text{ hours}$$
$$\approx 107 \text{ minutes}$$

b time in mountains $\approx (3 - 1.7778)$ hours
$$\approx 1.2222 \text{ hours}$$
distance = average speed $\times$ time
$$\approx 25 \text{ km h}^{-1} \times 1.2222 \text{ hours}$$
$$\approx 30.6 \text{ km}$$

28 a Since the series is arithmetic,
$$x - 27 = 42 - x$$
$$\therefore \quad 2x = 69$$
$$\therefore \quad x = 34.5$$
The common difference $d = x - 27 = 7.5$

b $$u_n = u_1 + (n-1)d$$
$\therefore$ since $u_5 = 27$, $27 = u_1 + 4d$
$$\therefore \quad u_1 = 27 - 4 \times 7.5$$
$$\therefore \quad u_1 = -3$$

c $$u_n = -3 + (n-1)7.5$$
$$= 7.5n - 10.5$$
If $7.5n - 10.5 = 2400$
then $7.5n = 2410.5$
$$\therefore \quad n = 321.4$$
So, the first term greater than 2400 is the 322nd term, and this is $u_{322} = -3 + 321 \times 7.5$
$$= 2404.5$$

29 a Since the sequence is geometric,
$$\dfrac{x}{45} = \dfrac{281.25}{x}$$
$$\therefore \quad x^2 = 45 \times 281.25$$
$$\therefore \quad x = \sqrt{12\,656.25} \quad \{\text{since } x > 0\}$$
$$\therefore \quad x = 112.5$$
So, the common ratio $r = \dfrac{112.5}{45} = 2.5$

b For a geometric series, $S_n = \dfrac{u_1(r^n - 1)}{r - 1}$
So, $u_5 + u_6 + \dots + u_{12}$
$$= S_{12} - S_4$$
$$= \dfrac{45(2.5^{12} - 1)}{11} - \dfrac{45(2.5^4 - 1)}{3}$$
$$\approx 243\,000$$

30 a Error $= 4 \text{ m} - 3.94 \text{ m}$
$$= 0.06 \text{ m}$$

b Length of the joined pipes $= 5 \times 3.94 \text{ m}$
$$= 19.7 \text{ m}$$

c The approximated length would be $5 \times 4 \text{ m} = 20 \text{ m}$
$\therefore$ the error $= 20 \text{ m} - 19.7 \text{ m}$
$$= 0.3 \text{ m}$$

d The percentage error $= \dfrac{|0.3|}{19.7} \times 100\%$
$$\approx 1.52\%$$

31 a $\dfrac{4.5}{5} = \dfrac{9}{10}$ and $\dfrac{4.05}{4.5} = \dfrac{9}{10}$
$\therefore$ the sequence is geometric with common ratio $\dfrac{9}{10}$.

b The third bounce has height $4.05 \times \dfrac{9}{10} = 3.645$ m
$$\approx 3.65 \text{ m}$$

c The total distance travelled
$$= 5 + 2 \times 4.5 + 2 \times 4.05 + 2 \times 3.645$$
$$\approx 29.4 \text{ m}$$

32 **a** **i** $A = \{-1, 0, 1, 2\}$ **ii** $B = \{2, 3, 5, 7, 11, 13\}$
  iii $C = \{-\sqrt{8}, +\sqrt{8}\}$
  b **i** false   **ii** true   **iii** true

33 **a** Commission $= 1.5\%$ of $60\,000$ JPY $= 900$ JPY
  b $59\,100$ JPY $= 59\,100 \times 0.010\,334$ AUD
  $= 610.74$ AUD
  So, 610.74 AUD would be received.
  c Commission $= 1.5\%$ of 610.74 AUD $= 9.16$ AUD
  Amount remaining $= (610.74 - 9.16)$ AUD
  $= 601.58$ AUD
  601.58 AUD $= 601.58 \times \dfrac{1}{0.010\,334}$ JPY
  $= 58\,214$ JPY
  So, there would be $58\,214$ JPY.

34 **a** The value after 5 years $= PV\left(1 + \frac{r}{100k}\right)^{kn}$
  where $PV = €40\,000$, $r =$ the interest rate
  $k =$ the number of compounds per year
  $n = 5$ years
  For $r = 5.05$ and $k = 12$,
  the final value $= €40\,000\left(1 + \frac{5.05}{1200}\right)^{60}$
  $= €51\,462.31$
  For $r = 5.2$ and $k = 1$,
  the final value $= €40\,000\left(1 + \frac{5.2}{100}\right)^{5}$
  $= €51\,539.32$
  So, more interest is earned at 5.2% p.a. compounding yearly.
  b The difference in interest is $€(11\,539.32 - 11\,462.31)$
  $= €77.01$

35 **a** For depreciation of 12.5% each year, we multiply the value by 0.875 at the end of each year. After n years,
  final value $=$ initial value $\times 0.875^n$
  $\therefore$ $\$21\,400 =$ initial value $\times 0.875^3$
  $\therefore$ initial value $= \dfrac{\$21\,400}{0.875^3}$
  $= \$31\,944.02$
  $\approx \$31\,900$
  b If the final value is $\$10\,000$ then
  $10\,000 = 31\,944.02 \times 0.875^n$
  Using technology, $n \approx 8.70$
  The value will fall below $\$10\,000$ in the 9th year.

36 **a** **i** There are $\frac{1}{2} \times 128 = 64$ clubs in the second round.
  ii There are $\frac{1}{2} \times 64 = 32$ clubs in the third round.
  b Let u_n be the number of teams in the nth round.
  The sequence is geometric with $u_1 = 128$ and $r = \frac{1}{2}$
  $\therefore u_n = 128 \times \left(\frac{1}{2}\right)^{n-1}$
  c If $u_n = 1$ then
  $128 \times \left(\frac{1}{2}\right)^{n-1} = 1$
  $2^{n-1} = 128$
  $\therefore 2^{n-1} = 2^7$
  $\therefore n - 1 = 7$ {equating indices}
  $\therefore n = 8$
  $\therefore$ in the 8th round there would be only 1 team, so it takes 7 rounds to determine the winner.

37 **a** Let the price of a book be b NZD and the price of a pen be p NZD.
  So, $15b + 8p = 209$ .... (1)
  and $7b + 3p = 96.25$ .... (2)
  b Solving (1) and (2) simultaneously using technology, $b = 13$ and $p = 1.75$.
  c The total cost of 10 books and 6 pens
  $= 10 \times 13 + 6 \times 1.75$
  $= 140.50$ NZD

38 **a** The series is geometric with $u_2 = 14.5$, and $u_5 = 1.8125$.
  Now $u_2 = u_1 \times r$ and $u_5 = u_1 \times r^4$
  $\therefore \dfrac{u_5}{u_2} = \dfrac{u_1 r^4}{u_1 r} = r^3$
  $\therefore r^3 = \dfrac{1.8125}{14.5} = 0.125$
  $\therefore r = 0.5$
  The common ratio is 0.5.
  b $u_2 = u_1 \times r$
  $\therefore u_1 = \dfrac{u_2}{r} = \dfrac{14.5}{0.5} = 29$
  $\therefore$ the first term is 29.
  c For a geometric series, $S_n = \dfrac{u_1(1 - r^n)}{1 - r}$
  $\therefore S_5 = \dfrac{29(1 - 0.5^5)}{1 - 0.5}$
  $\therefore S_5 = 56.1875$

39 **a** Each year the population is multiplied by 1.09
  $\therefore$ the population in 2015 is $1200 \times 1.09^5 \approx 1850$ people.
  b If n is the number of years after 2010, then
  $1200 \times 1.09^n = 2500$
  $\therefore 1.09^n = \frac{25}{12}$
  Using technology, $n \approx 8.52$
  $\therefore$ the population reaches 2500 during the year 2019.
  c If the population increased with multiplier r each year,
  then $1200 \times r^{10} = 3200$
  $\therefore r^{10} = \frac{32}{12} = \frac{8}{3}$
  Using technology, $r \approx 1.103$
  $\therefore$ the rate of increase $\approx 10.3\%$

40 **a** The series is geometric if $\dfrac{k - 6}{k} = \dfrac{k}{2k - 9}$
  $\therefore (k - 6)(2k - 9) = k^2$
  $\therefore 2k^2 - 9k - 12k + 54 = k^2$
  $\therefore k^2 - 21k + 54 = 0$
  $\therefore (k - 3)(k - 18) = 0$
  $\therefore k = 3$ or 18
  b The sequence is arithmetic if
  $(k - 6) - k = k - (2k - 9)$
  $\therefore -6 = -k + 9$
  $\therefore k = 15$

41 **a** **i** $s = 16$, $r = 5$ **ii** $u_1 = 27$, $d = 5$
  b $a + 12p = 20$ .... (1)
  $a + 22p = 34$ .... (2)
  Using technology, $a = 3.20$ and $p = 1.40$.

42 a Using $FV = PV\left(1 + \dfrac{r}{100}\right)^n$,

$$1200 = 6000 \times \left(1 + \frac{r}{100}\right)^5$$

$\therefore \ r \approx -27.522$ {using technology}

So, the rate of depreciation is 27.5% per annum.

b At the end of each year we multiply the value by
$1 - 0.275\,22 = 0.724\,78$

So, the values are €4348.68, €3151.84, €2284.39,
€1655.68, then €1200.

43 a
$$\text{The interest}\ \ I = PV\left(1 + \frac{r}{100k}\right)^{kn} - PV$$

$$\therefore \ \ 2000 = 6500\left(1 + \frac{8}{400}\right)^{4n} - 6500$$

$$\therefore \ \ \left(1 + \frac{8}{400}\right)^{4n} = \frac{8500}{6500}$$

Using technology, $n \approx 3.39$ years

$\therefore$ it will take 14 quarters to earn \$2000 in interest.

b In 14 quarters the actual interest will be
$$I = 6500\left(1 + \frac{8}{400}\right)^{14} - 6500$$
$$= \$2076.61$$

To earn this in interest compounding monthly,
$$2076.61 = 6500\left(1 + \frac{r}{1200}\right)^{42} - 6500$$

Using technology, $r \approx 7.947$

The rate required is 7.95% p.a.

44 a
$$178 - 4n = 7n + 57$$
$$\therefore \ \ 11n = 121$$
$$\therefore \ \ n = 11$$

The sequences have the 11th term in common, and this
term is $178 - 4 \times 11 = 134$.

b
$$C = R \ \text{if}\ 25n = 21\,000 + 7.5n$$
$$\therefore \ \ 17.5n = 21\,000$$
$$\therefore \ \ n = 1200$$

The cost equals the revenue when 1200 goods are produced
and sold.

c If $\dfrac{n(n+1)}{2} = 435$

$$\text{then}\ \ n(n+1) = 870$$
$$\therefore \ \ n^2 + n - 870 = 0$$
$$\therefore \ \ (n + 30)(n - 29) = 0$$
$$\therefore \ \ n = 29 \ \ \text{\{since}\ n > 0\}$$

The sum exceeds 435 for $n > 29$.

45 a $u_7 = 4374\left(\frac{1}{3}\right)^6$
$= 6$

b If $4374\left(\frac{1}{3}\right)^{n-1} = 0.1$
$$\text{then}\ \ 3^{n-1} = 43\,740$$
Using technology, $n \approx 10.7$
So, $u_n \leqslant 0.1$ for $n \geqslant 11$.

c For a geometric series, $S_n = \dfrac{u_1(1 - r^n)}{1 - r}$

$$\therefore \ \ S_8 = \frac{4374\left(1 - (\frac{1}{3})^8\right)}{1 - \frac{1}{3}}$$

$$\therefore \ \ S_8 = 6560$$

46 a Let the length be l cm $\therefore \ 2x + 2l = 80$
$$\therefore \ \ 2l = 80 - 2x$$
$$\therefore \ \ l = 40 - x$$

b The area $A = l \times x$
$$= x(40 - x)$$
$$= 40x - x^2 \ \text{cm}^2$$

c
$$40x - x^2 = 375$$
$$\therefore \ \ x^2 - 40x + 375 = 0$$
$$\therefore \ \ (x - 15)(x - 25) = 0$$
$$\therefore \ \ x = 15 \ \text{or}\ 25$$
$$\therefore \ \ l = 25 \ \text{or}\ 15$$

So, the rectangle is 15 cm × 25 cm.

47 a Book value of computer $= 4500\left(1 - \frac{22.5}{100}\right)^3$
$$= 2094.68 \ \text{CAD}$$

b i If the original cost of the tractor was £PV, then
$$29\,000 = PV \times \left(1 - \frac{17.5}{100}\right)^5$$
$$\therefore \ \ \frac{29\,000}{\left(1 - \frac{17.5}{100}\right)^5} = PV$$
$$\therefore \ \ PV \approx 75\,880$$

So, the original cost of the tractor was £75 880.

ii The total depreciation for the 5 years is
$$£(75\,880 - 29\,000) = £46\,880.$$

iii The lease value per year must be
$$£\frac{46\,880}{5} = £9376$$

48 a $s = 0$ where $ut - 5t^2 = 0$
$$\therefore \ \ 70t - 5t^2 = 0$$
$$\therefore \ \ -5t(t - 14) = 0$$
$$\therefore \ \ t = 0 \ \text{or}\ 14$$

The rocket is in the air for 14 seconds.

b $s = 30$ when $ut - 5t^2 = 30$
$$\therefore \ \ 70t - 5t^2 = 30$$
$$\therefore \ \ 5t^2 - 70t + 30 = 0$$
Using technology, $t \approx 0.443$ or 13.557

The rocket is above 30 m for
$$13.557 - 0.443 \approx 13.1 \ \text{seconds}$$

49
$$23r + b = 19.15 \quad \dots \ (1)$$
$$15r + b = 15.95 \quad \dots \ (2)$$

Solving (1) and (2) simultaneously using technology,
$r = 0.4$ and $b = 9.95$.

LONG QUESTIONS

1 a i $u_1 = 4 + 11 \times 1 = 15$
$u_2 = 4 + 11 \times 2 = 26$

ii For an arithmetic sequence,
$$S_n = \frac{n}{2}(2u_1 + (n - 1)d)$$

In this case $u_1 = 15$ and $d = 11$
$$\therefore \ \ S_{10} = \frac{10}{2}(30 + 9 \times 11)$$
$$= 645$$

b i $u_5 = 4(2.2)^4 = 93.7024$ {exact}

ii For a geometric sequence, $S_n = \dfrac{u_1(r^n - 1)}{r - 1}$

In this case $u_1 = 4$ and $r = 2.2$
$$\therefore \ \ S_{10} = \frac{4(2.2^{10} - 1)}{2.2 - 1} \approx 8850$$

c For the arithmetic sequence: $u_1 = 15$, $u_2 = 26$, $u_3 = 37$, $u_4 = 48$, $u_5 = 59$

For the geometric sequence: $u_1 = 4$, $u_2 = 8.8$, $u_3 = 19.36$, $u_4 = 42.592$, $u_5 = 93.7024$

$\therefore$ the fifth term of geometric sequence is greater than that of the arithmetic sequence. The difference between these terms is ≈ 34.7.

2 a

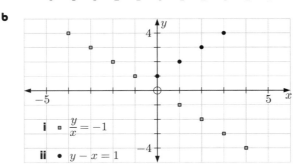

b

    i $\square$ $\dfrac{y}{x} = -1$

    ii $\bullet$ $y - x = 1$

c Let $x = -2$, $y = -3$, so $x^2 = 4$ and $y^2 = 9$.
We see that $x > y$ but $x^2 < y^2$.

3 a
$$u_5 = u_1 + 4d$$
$$\therefore \quad u_1 + 4d = 51 \quad .... \ (1)$$
$$S_n = \frac{n}{2}(2u_1 + (n-1)d)$$
$$\therefore \quad nu_1 + \frac{n(n-1)}{2}d = S_n$$
$$\therefore \quad 5u_1 + 10d = 185 \quad \{S_5 = 185\}$$
$$\therefore \quad u_1 + 2d = 37 \quad .... \ (2)$$

b Using technology, $d = 7$ and $u_1 = 23$

c $u_n = 23 + (n-1)7$
$$= 7n + 16$$

If $7n + 16 = 1000$ then $7n = 984$
$$\therefore \quad n \approx 140.6$$

$\therefore$ the first term to exceed 1000 is the 141st term, and this is $u_{141} = 7 \times 141 + 16$
$$= 1003$$

d **i** $S_n = \dfrac{n}{2}(2u_1 + (n-1)d)$
$$\therefore \quad \frac{k}{2}(2 \times 23 + (k-1)7) = 3735$$
$$\therefore \quad k(46 + 7k - 7) = 7470$$
$$\therefore \quad 7k^2 + 39k - 7470 = 0 \quad \text{as required}$$

  ii Using technology, $k = -\frac{249}{7}$ or 30
But $k > 0$, so $k = 30$.

4 a **i** false since '2' is not included in $\{x \mid x < 2, \ x \in \mathbb{R}\}$

  ii true   **iii** false since $2^2 + 2 = 6 \neq 2$

b **i** $A = \{14, 21, 28\}$, $B = \{14, 28\}$,
$C = \{20, 22, 24, 26, 28\}$.

  ii

c **B** is false. Consider $p = 1$, $q = 2$. $1 - 2 \neq 2 - 1$

5 a $25.32 \times \dfrac{6.057}{2.4 \times \sqrt{5.14}} \approx 28.185\,68$

  i 28.186   {to 5 significant figures}

  ii 28.2   {to the nearest tenth}

  iii 30   {to 1 significant figure}

b **i** The actual length covered is $3 \times 3.63 = 10.89$ m.

  ii The error between the actual length and the stated length is $10.80 - 10.89 = -0.09$ m.

Percentage error $= \dfrac{|10.80 - 10.89|}{10.89} \times 100\%$
$$= 0.826\%$$

c **i** The minimum volume needed is
$10.89 \times 10.89 \times 0.095 \approx 11.266 \text{ m}^3 \approx 11.3 \text{ m}^3$
Maximum volume needed is
$10.89 \times 10.89 \times 0.105 \approx 12.452 \text{ m}^3 \approx 12.5 \text{ m}^3$

  ii Planned volume $= 10.8 \times 10.8 \times 0.1$
$$= 11.664 \text{ m}^3$$
The difference in cost is
$$(12.452 - 11.664) \times 47.50$$
$$= \text{€}37.43$$

  iii The percentage difference in cost is
$$\frac{37.43}{11.664 \times 47.50} \times 100\% = 6.76\%$$

6 a **i** 120, 123, 126   **ii** 4, 11, 18

b The second sequence $u_{n+1} = u_n + 7$ is arithmetic with $u_1 = 4$ and common difference $d = 7$
$$\therefore \quad u_n = 4 + 7(n-1)$$
The sequences have a term in common when
$$120 + 3(n-1) = 4 + 7(n-1)$$
$$\therefore \quad 117 + 3n = -3 + 7n$$
$$\therefore \quad 4n = 120$$
$$\therefore \quad n = 30$$
So, the 30th terms are the same, and they have value
$$u_{30} = 4 + 7 \times 29$$
$$= 207$$

c If $120 + 3(n-1) = 151$
then $3n + 117 = 151$
$$\therefore \quad 3n = 34$$
$$\therefore \quad n \approx 11.3 \text{ which is not an integer.}$$
If $4 + 7(n-1) = 151$
then $7n - 3 = 151$
$$\therefore \quad 7n = 154$$
$$\therefore \quad n = 22$$
So, 151 is the 22nd term of the second sequence.

d For an arithmetic sequence,
$$S_n = \frac{n}{2}(2u_1 + (n-1)d)$$
$$\therefore \quad \frac{n}{2}(2 \times 120 + (n-1)3) = \frac{n}{2}(2 \times 4 + (n-1)7)$$
$$\therefore \quad \frac{n}{2}(240 + 3n - 3 - 8 - 7n + 7) = 0$$
$$\therefore \quad \frac{n}{2}(-4n + 236) = 0$$
$$\therefore \quad -2n(n - 59) = 0$$
$$\therefore \quad n = 59$$
$$\{\text{since } n > 0\}$$
The sums of the first 59 terms are the same.

e
$$\frac{n}{2}(2 \times 120 + (n-1)3)$$
$$-\frac{n}{2}(2 \times 4 + (n-1)7) = 228$$
$$\therefore \quad \frac{n}{2}(-4n + 236) - 228 = 0$$
$$\therefore \quad -2n^2 + 118n - 228 = 0$$
$$\therefore \quad n^2 - 59n + 114 = 0$$
$$\therefore \quad (n-57)(n-2) = 0$$
$$\therefore \quad n = 2 \text{ or } 57$$

The sum of the first sequence exceeds the sum of the second by 228 for both the 2nd and 57th terms.

7 a $p = \dfrac{1}{0.7} \approx 1.429, \quad q = \dfrac{1}{1.085} \approx 0.922, \quad r = 1.000$

b Dean will receive $\quad 20\,000 \times 0.700 = 14\,000$ GBP

c Dean will receive $\quad 8000 \times 1.550 = 12\,400$ USD

d Commission = 2% of \$2500
$$= \$50$$

So, Dean actually receives $\quad 2450 \times 1.085$
$$= 2658.25 \text{ CHF}$$
$$\approx 2658 \text{ CHF}$$

e $FV = PV\left(1 + \dfrac{r}{100k}\right)^{kn}$ where $r = 5.1$ and $k = 12$.
kn is the number of months.
The investment is doubled when
$$\left(1 + \frac{r}{100k}\right)^{kn} = 2$$
$$\therefore \quad \left(1 + \frac{5.1}{1200}\right)^{kn} = 2$$
$$\therefore \quad kn \approx 163.4 \quad \{\text{using technology}\}$$
The investment is doubled after 164 months.

8 a Total surface area $= $ length $\times$ width
$$= (2x+8)(x-2) \text{ m}^2$$
$$= 2x^2 - 4x + 8x - 16 \text{ m}^2$$
$$= 2x^2 + 4x - 16 \text{ m}^2$$

b
$$2x^2 + 4x - 16 = 224$$
$$\therefore \quad 2x^2 + 4x - 240 = 0$$
$$\therefore \quad x^2 + 2x - 120 = 0$$
$$\therefore \quad (x+12)(x-10) = 0$$
$$\therefore \quad x = -12 \text{ or } 10$$
But $x > 0$, so $x = 10$

c Total capacity
$= $ capacity of swimming pool + capacity of wading pool
$= 1.8 \times (x-2) \times (x+8) + 0.5 \times (x-2) \times x$ kL
$= 1.8 \times 8 \times 18 + 0.5 \times 8 \times 10$ kL
$= 299.2$ kL
$= 299\,200$ L

d i Original capacity of wading pool
$$= 0.5 \times 8 \times 10 \text{ kL}$$
$$= 40 \text{ kL}$$
$$= 40\,000 \text{ L}$$
$$\therefore \quad \text{new capacity of wading pool}$$
$$= 40\,000 - 2400$$
$$= 37\,600 \text{ L}$$

ii Let the new depth of the wading pool be d m.
$$\therefore \quad d \times 8 \times 10 = 37.6$$
$$\therefore \quad d = 0.47 \text{ m}$$
So, the new depth of the wading pool is 47 cm.

9 a $a = 1 + \dfrac{7.5}{400} = 1.018\,75$
$b = 1 + \dfrac{10}{1200} \approx 1.008\,33$
$c = 1 + \dfrac{12.5}{100} = 1.125$

b i $FV = PV\left(1 + \dfrac{r}{100k}\right)^{kn}$
$$= 4000 \times 1.006\,25^{5 \times 12}$$
$$= 5813.18 \text{ CHF}$$

ii If $5000 = 4000 \times 1.025^{kn}$
then $kn \approx 9.04$
So, it will take 10 quarters.

c Using the table, $10\,000 = PV \times (1.010\,417)^{7 \times 12}$
$$\therefore \quad PV = \frac{10\,000}{1.010\,417^{84}} \approx 4187.41$$
The amount required is 4187.41 CHF.

d i Commission $= 1.5\%$ of \$6700 $= \$100.50$
The amount received $= 6599.5 \times 1.2778$
$$= 8432.84 \text{ CHF}$$

ii $FV = PV\left(1 + \dfrac{r}{100k}\right)^{kn}$
$$= 8432.84(1.015)^3$$
$$= 8818.04 \text{ CHF}$$

e Commission $= 1.5\%$ of 8818.04 CHF $= 132.27$ CHF.
The return is $\dfrac{8685.77}{1.2778} = 6797.44$ USD.

f We assume that the exchange rate has remained constant.

10 a i $\dfrac{u_2}{u_1} = \dfrac{28}{56} = 0.5, \qquad \dfrac{u_3}{u_2} = \dfrac{14}{28} = 0.5$
Since $\dfrac{u_2}{u_1} = \dfrac{u_3}{u_2}$, the sequence is geometric.

ii $u_n = u_1 r^{n-1}$
$$\therefore \quad u_8 = 56 \times 0.5^7 = 0.4375$$

iii $S_n = \dfrac{u_1(1 - r^n)}{1 - r}$
$$\therefore \quad S_8 = \frac{56 \times (1 - 0.5^8)}{1 - 0.5} = 111.5625$$

b i $u_3 = u_1 r^2 = 24.5$
$u_5 = u_1 r^4 = 12.005$
$$\therefore \quad \frac{u_5}{u_3} = r^2 = \frac{12.005}{24.5} = 0.49$$
$$\therefore \quad r = \pm 0.7$$
But all of the terms are positive, so $r > 0$
$$\therefore \quad r = 0.7$$
Now, $u_1 \times 0.49 = 24.5$
$$\therefore \quad u_1 = 50$$
So, the first term is 50 and the common ratio is 0.7.

ii The general term $u_n = 50 \times 0.7^{n-1}$

c If $20 \times 0.8^{n-1} = 50 \times 0.7^{n-1}$
then $\left(\dfrac{0.8}{0.7}\right)^{n-1} = \dfrac{50}{20} = 2.5$
Using technology, $n \approx 7.86$
So, the first 7 terms of the sequence in **b** are larger than the first 7 terms of $u_n = 20 \times 0.8^{n-1}$.
$$\therefore \quad n = 7$$

d $S_n = \dfrac{u_1 \times (1 - r^n)}{1 - r}$

   i $S_{30} = \dfrac{20 \times (1 - 0.8^{30})}{1 - 0.8} \approx 99.9$

   ii $S_{50} = \dfrac{20 \times (1 - 0.8^{50})}{1 - 0.8} \approx 100$

   iii $S_{100} = \dfrac{20 \times (1 - 0.8^{100})}{1 - 0.8} \approx 100$

For large values of n, the sum approaches 100.

11 a i At the end of the year, 12.5% interest is added, and then \$$k$ is paid off.
12.5% interest has multiplier 1.125, so the amount owing is $\$(20\,000 \times 1.125 - k)$.

   ii At the end of the second year, the amount owing is $\$((20\,000 \times 1.125 - k) \times 1.125 - k)$.

   iii Using the expression in **ii**,
$$25\,312.5 - 1.125k - k = 17\,131.25$$
$$\therefore \quad 2.125k = 8181.25$$
$$\therefore \quad k = 3850$$

b i percentage decrease $= \dfrac{\text{decrease}}{\text{original value}} \times 100\%$
$$= \dfrac{24\,000 - 20\,400}{24\,000} \times 100\%$$
$$= 15\%$$

   ii Depreciating at 15%, the value after two years
$$= \$20\,400 \times 0.85$$
$$= \$17\,340$$

   iii The value after n years $= \$24\,000 \times (0.85)^n$

   iv

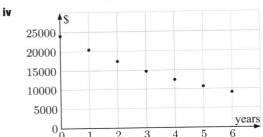

12 a i $A_1 = 6$ **ii** Common difference $d = 19$

b i $G_n = G_1 \times r^{n-1}$
$$\therefore \quad G_3 = G_1 \times r^2 = 1920$$
and $G_{10} = G_1 \times r^9 = 15$
$$\therefore \quad \dfrac{G_{10}}{G_3} = r^7 = \dfrac{15}{1920}$$
$$\therefore \quad r^7 = \dfrac{1}{128} = \left(\dfrac{1}{2}\right)^7$$
$$\therefore \quad r = \dfrac{1}{2}$$
The common ratio is $\dfrac{1}{2}$.

   ii $G_1 \times \left(\dfrac{1}{2}\right)^2 = 1920$
$$\therefore \quad G_1 = 7680$$

c If $A_n = G_n$ then $19n - 13 = 7680 \times \left(\dfrac{1}{2}\right)^{n-1}$
Using technology, $n = 7$
The 7th term is common, and this is $A_7 = G_7 = 120$

d i A_n is arithmetic, so $S_{A_n} = \dfrac{n}{2}(2A_1 + (n-1)d)$
$$= \dfrac{n}{2}(12 + (n-1)19)$$
$$= \dfrac{19}{2}n^2 - \dfrac{7}{2}n$$

   ii G_n is geometric, so $S_{G_n} = \dfrac{G_1(1 - r^n)}{1 - r}$
$$= \dfrac{7680(1 - (\frac{1}{2})^n)}{1 - \frac{1}{2}}$$
$$= 15\,360(1 - (\tfrac{1}{2})^n)$$

   iii If $\frac{19}{2}n^2 - \frac{7}{2}n = 15\,360\left(1 - (\frac{1}{2})^n\right)$
Using technology, $n = 0$ or $n \approx 40.4$
We know $n > 0$, so $S_{A_n} < S_{G_n}$ for $1 \leqslant n \leqslant 40$
It takes 41 terms for S_{A_n} to become greater than S_{G_n}.

13 a i Eddie's savings after 3 years will be
$$FV = PV\left(1 + \tfrac{r}{100k}\right)^{kn}$$
$$= 1800\left(1 + \tfrac{5.6}{400}\right)^{4 \times 3}$$
$$= \pounds 2126.81$$

   ii The additional amount to reach the target is
$$\pounds(5500 - 2126.81) = \pounds 3373.19$$

b Let PV be the amount needed to accumulate to £5500.
Using $FV = PV\left(1 + \tfrac{r}{100k}\right)^{kn}$
$$5500 = PV \times \left(1 + \tfrac{5.6}{400}\right)^{4 \times 3}$$
$$\therefore \quad \dfrac{5500}{\left(1 + \tfrac{5.6}{400}\right)^{12}} = PV$$
$$\therefore \quad PV = \pounds 4654.87$$
Eddie's father needs to add
$$\pounds(4654.87 - 1800) = \pounds 2854.87$$

c i Interest $I = \dfrac{PV \times rn}{100}$
$$= \dfrac{5500 \times 9 \times 2}{100}$$
$$= \pounds 990$$
$$\therefore \quad \text{total to repay} = \pounds 6490$$

   ii The monthly repayment $= \pounds 6490 \div 24 = \pounds 270.42$.

d Thomas pays $\pounds(6490 - 4654.87) = \pounds 1835.13$ more.

14 a i   Interest $= \$5000 \times 0.012 = \$60$
$$\therefore \quad \text{repayment} = \$250 + \$60 = \$310$$

   ii New principal $= \$5000 + \$60 - \$310$
$$= \$4750$$
$$\text{Interest} = \$4750 \times 0.012 = \$57$$
$$\therefore \quad \text{repayment} = \$250 + \$57 = \$307$$

   iii New principal $= \$4750 + \$57 - \$307$
$$= \$4500$$
$$\text{Interest} = \$4500 \times 0.012 = \$54$$
$$\therefore \quad \text{repayment} = \$250 + \$54 = \$304$$

b   First term $u_1 = \$310$
Common difference $d = -\$3$
The nth repayment $u_n = \$(310 - 3(n-1))$
$$= \$(313 - 3n)$$

c $u_{10} = \$(313 - 3 \times 10)$
$$= \$283$$

d Every month the principal reduces by \$250
$$\therefore \quad \text{it will take } \dfrac{5000}{250} = 20 \text{ monthly repayments.}$$

e For the arithmetic sequence, $S_n = \frac{n}{2}(2u_1 + (n-1)d)$

∴ the total amount repaid is
$$S_{20} = \$\frac{20}{2}(2 \times 310 + 19 \times (-3))$$
$$= \$10(620 - 57)$$
$$= \$5630$$

15 a i Weight at start of second year
$$= 200 \times 0.9 = 180 \text{ g}$$
Weight at start of third year
$$= 180 \times 0.9 = 162 \text{ g}$$

ii Common ratio is 0.9

iii Weight at start of sixth year
$$= 200 \times 0.9^5 = 118.098 \text{ g}$$

iv

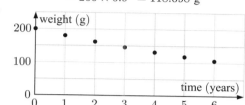

v Weight at start of nth year $= 200 \times 0.9^{n-1}$
If $200 \times 0.9^{n-1} = 20$ then $0.9^{n-1} = \frac{1}{10}$
Using technology, $n \approx 22.9$
So, the material will weigh less than 20 g at the start of the 23rd year.

b The amount of radioactive material at the beginning of the nth year is $120r^{n-1}$.

At the end of the 6th year, or the beginning of the 7th year, the amount is
$$120 \times r^{7-1} = 49.152$$
$$\therefore r^6 = \frac{49.152}{120}$$
$$\therefore r = \left(\frac{49.152}{120}\right)^{\frac{1}{6}} \approx 0.862$$

The annual decrease is
$$\approx 1 - 0.862 \approx 0.138 \approx 13.8\%.$$

SOLUTIONS TO TOPIC 2 (DESCRIPTIVE STATISTICS)

SHORT QUESTIONS

1 a Discrete data.

b Ordered data set: 14, 16, 18, 23, 24, 25, 26, 26, 34

Mean $= \dfrac{14 + 16 + ... + 34}{9} \approx 22.9$ customers

The median is the $\left(\dfrac{n+1}{2}\right) = \dfrac{9+1}{2} = 5$th value

∴ the median $= 24$ customers
Mode $= 26$ customers
Range $= 34 - 14 = 20$ customers

c Total income $= 206 \times \$14.20 = \$2925.20$

2 a Discrete data.

b

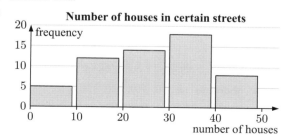

c Modal class is 30 - 39 houses

3 a The ordered data is:

3	5	7	13	14	17	17	18	20	22	22
22	25	27	31	34	35	36	39	40	42	49

$$\left(\frac{n+1}{2}\right) = \frac{22+1}{2} = 11.5 \text{ th value}$$

The 11th and 12th values are both 22.
∴ the median is 22 birds per day.
Q_1 is the 6th value.
∴ $Q_1 = 17$ birds per day
Q_3 is the 17th value.
∴ $Q_3 = 35$ birds per day

b On 5 of the 22 days there were more than 35 birds in the park.

∴ P(there are more than 35 birds) $= \frac{5}{22}$.

4 a

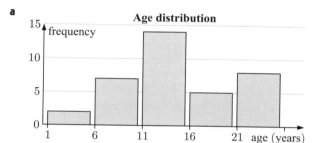

b

Age (years)	Frequency	Midpoint
1 - 5	2	3
6 - 10	7	8
11 - 15	14	13
16 - 20	5	18
21 - 25	8	23

Using technology, the mean ≈ 14.4 years.

c The modal class is 11 - 15 years.

5 a $\dfrac{90 + 100 + 93 + 96 + p + 107 + 98 + 98 + 92}{9} = 97$
$$\therefore 774 + p = 873$$
$$\therefore p = 99$$

b The ordered data is:
90, 92, 93, 96, 98, 98, 99, 100, 107
∴ the median is 98.

6 a The modal class is $4 \leqslant x < 6$.

b

Score	Frequency	Midpoint
$0 \leqslant x < 2$	2	1
$2 \leqslant x < 4$	7	3
$4 \leqslant x < 6$	15	5
$6 \leqslant x < 8$	11	7
$8 \leqslant x < 10$	5	9

Using technology, the mean ≈ 5.5.

c

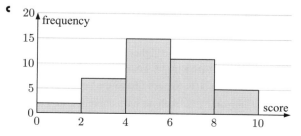

7 a The data is continuous.

b Mid-interval values are: 5, 15, 25, 35, 45

c Mean $= \dfrac{(5 \times 2) + (15 \times 15) + ... + (45 \times 5)}{50}$

$= 24.6$

d Cumulative frequencies: 2, 17, 38, 45, 50.

e

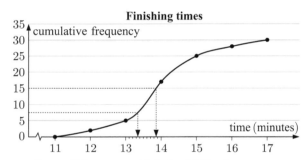

Height of plants

f The 80th percentile is approximately 32 cm.

g The 80th percentile is the 40th plant. There are 10 plants taller than the 80th percentile.

8 a number of houses is $3 + 17 + 12 + 5 + 3 = 40$

b number of people is
$(3 \times 1) + (17 \times 2) + (12 \times 3) + (5 \times 4) + (3 \times 5) = 108$

c mean $= \frac{108}{40} = 2.7$ people per house.

d $\dfrac{12 + 5 + 3}{40} \times 100\% = 50\%$ of houses have more than 2 people.

9

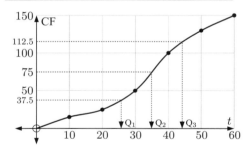

Finishing times

a median $\approx$ 15th score ≈ 13.8 (or 13.9) minutes

b $Q_1 \approx 7\frac{1}{2}$th score ≈ 13.3 min.
So, any time less than 13.3 minutes.

c 2 finish within 12 minutes and 25 finish within 15 min
$\therefore$ 23 finish between 12 and 15 minutes.

10 a 9, 10, a, 13, b, 16, 21
Sum $= 7 \times 14 = 98$
$\therefore$ $(a + b) = 98 - (9 + 10 + 13 + 16 + 21) = 29$
In order, $a = 13$, $b = 16$. {Any other choices of a between 10 and 13 or b between 13 and 16 are too small.}

b The median is the average of 9 and 11
and so the median $= 10$.
Since median $=$ mean, mean $= 10$.
$$\text{mean} = \frac{1 + 5 + 9 + 11 + 16 + p}{6}$$
$$\therefore \quad 10 = \frac{42 + p}{6}$$
$$\therefore \quad 42 + p = 60 \text{ and so } p = 18$$

11 a Company 2 has the higher standard deviation and therefore greater dispersion of wages.

b Company 1 is likely to have the smaller IQR.

c Company 2 is likely to have more highly paid employees. Data is more spread and the median wage is slightly higher.

12 a $165 \leqslant$ height < 175

b The frequency table is

Height (rounded, cm)	Frequency
140	2
150	6
160	14
170	17
180	9
190	2

Total number of students is 50.
Using technology, mean height ≈ 166 cm.
Standard deviation is 11.5 cm.

c $\overline{x} + 2s \approx 166 + 23 \approx 189$
There are two students in the range $185 \leqslant$ height < 195. These may or may not be taller than 189 cm, so 0, 1, or 2 students are taller than 2 standard deviations above the mean.

13 a mean ≈ 24.1, standard deviation ≈ 6.71

{using technology}

b The elimination bound is $\approx 24.1 + 6.71 \approx 30.81$ minutes. So, the contestants who took 32 minutes and 40 minutes are eliminated.

c mean $= 21.7$, standard deviation $= 4.1$

d Elimination bounds are $21.7 \pm 4.1 = 17.6$ to 25.8 minutes.
$\therefore$ 6 contestants (those taking 20, 25, 19, 21, 25, 22 minutes) will compete in the next game.

14 a

Time	f	Cumulative frequency
$0 \leqslant t < 10$	15	15
$10 \leqslant t < 20$	10	25
$20 \leqslant t < 30$	25	50
$30 \leqslant t < 40$	50	100
$40 \leqslant t < 50$	30	130
$50 \leqslant t < 60$	20	150

b **i** median ≈ 35 min $(= Q_2)$ **ii** IQR $= Q_3 - Q_1$
$\approx 44 - 25$
≈ 19

iii $\approx 50\%$ of the data is less than 35 min
$\therefore$ the probability ≈ 0.5

15 a **i** mean $= \dfrac{55 + 58 + 61 + ... + 110}{10} = 76$ s

ii standard deviation ≈ 17.2 s

b Let Greg's time be x seconds
$$\therefore \quad \frac{55 + 58 + 61 + ... + 110 + x}{11} = 75$$
$$\therefore \quad 760 + x = 825$$
$$\therefore \quad x = 65$$
Greg's running time is 65 seconds.

16 a Midpoints are 90, 110, 130, 150, 170, 190
Using technology, mean rent $\approx €138.99$
Standard deviation ≈ 21.6 euros

b $30 + 14 + 1 = 45$ houses have rent €140 or greater.
There are 89 houses in total.
$\therefore \;\; \text{P(rent} \geqslant €140) = \frac{45}{89}$

c $\overline{x} + 1s = €138.99 + €21.60 = €160.59 \approx €160$

Percentage of rent above €160 is $\qquad \dfrac{14 + 1}{89} \times 100\%$

$\approx 16.9\%$

17 **a** Site 3 has the greatest range.
b Site 2 has the smallest spread.
c Site 1 has the highest median weight.
d The heaviest fungi were found at site 3.
e Site 1 has the highest proportion of weights above 40 g.

18 **a** $\text{range} = \text{max} - \text{min}$
$\therefore \;\; 26 = 39 - p$
$\therefore \;\; p = 13$

b **i** $\qquad \text{mean} = \dfrac{13 + 15 + ... + 39}{12}$

$\therefore \;\; 23 = \dfrac{165 + 2q + 3r}{12}$

$\therefore \;\; 276 = 165 + 2q + 3r$

$\therefore \;\; 2q + 3r = 111 \;\; .... \;(1)$

Also, the median is the $\dfrac{\text{6th value} + \text{7th value}}{2}$

$\therefore \;\; 22 = \dfrac{q + r}{2}$

$\therefore \;\; q + r = 44 \;\; .... \;(2)$

ii $\qquad 2q + 3r = 111$
$\qquad \underline{-2q - 2r = -88} \;\; \{-2 \times (2)\}$
Adding, $\quad r = 23$
Substituting $r = 23$ into (2) gives $q = 21$.
So, $q = 21$ and $r = 23$

c $\text{mode} = 23$

19 **a** $Q_1 \approx 175, \;\; Q_2 \approx 190, \;\; Q_3 \approx 200$
b $\text{Range} \approx 220 - 130 \approx 90$
c $\text{IQR} \approx 200 - 175 \approx 25$

20 **a** In order, the scores are:
25   48   49   57   59   61   62   67   68   70
72   75   75   76   78   78   81   82   85   87
$\text{min} = 25, \;\; Q_1 = 60, \;\; Q_2 = 71, \;\; Q_3 = 78, \;\; \text{max} = 87$

b
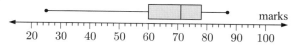

21 **a**
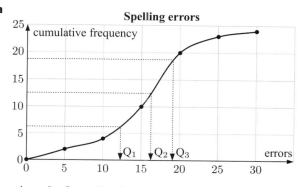

$\text{min} = 0, \;\; Q_1 \approx 12, \;\; Q_2 \approx 16, \;\; Q_3 \approx 19, \;\; \text{max} = 30$

b
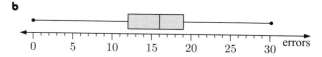

22 **a** Median is 5th value $= £310\,000$
b Q_1 is between 2nd and 3rd value $= £261\,000$
Q_3 is between 7th and 8th value $= £335\,000$
$\text{IQR} = £335\,000 - £261\,000 = £74\,000$
c Without the score 760, the new median price is £295 000.
The percentage change is
$$\frac{310\,000 - 295\,000}{310\,000} \times 100\% \approx 4.84\%$$

23 **a** discrete data
b $\text{mode} = 1$ car
c $78 + 117 + 69 + 18 + 2 = 284$ families surveyed
d **i** mean ≈ 1.12 cars
ii standard deviation ≈ 0.906 cars

24 **a** **i** A is the minimum value of X.
ii B is Q_1, the lower quartile.
iii C is Q_2, the median.
iv D is Q_3, the upper quartile.
v E is the maximum value of X.
b **i** $E - A = \text{maximum value} - \text{minimum value}$ is the range.
ii $D - B = Q_3 - Q_1$ is the interquartile range or IQR.
c **i** 75% of the scores are less than D, and 25% of the scores are less than B.
$\therefore \;\; \text{P}(B \leqslant X \leqslant D) = 0.75 - 0.25 = 0.5$
ii 75% of the scores are greater than B.
$\therefore \;\; \text{P}(X \geqslant B) = 0.75$

25 **a**

Minimum	Q$_1$	Median	Q$_3$	Maximum
1	3	4.5	5	7

b $\text{IQR} = Q_3 - Q_1$
$= 5 - 3$
$= 2$

c Histogram **A** could represent the data, as it is positively skewed.

26 **a** This is continuous numerical data.
b Average time $= \dfrac{12.1 + 13.3 + ... + 12.4}{20}$
≈ 11.9 minutes
c Standard deviation is a measure of the spread or dispersion of a data set.
d Using technology, $\sigma \approx 1.58$

27 The mode $= 19$, so $p = 19$

The median is 22.5, so $\dfrac{(20 + q)}{2} = 22.5$
$\therefore \;\; q = 25$

The average is 24.
$\therefore \;\; \dfrac{15 + 17 + 19 + 19 + 19 + 20 + 25 + 26 + 29 + 30 + 30 + r}{12}$
$= 24$

$\therefore \;\; \dfrac{249 + r}{12} = 24$
$\therefore \;\; r = 24 \times 12 - 249$
$\therefore \;\; r = 39$

28 a

	Min.	Q$_1$	Median	Q$_3$	Max.
New batteries	19	23.5	26	28	31

b Calculating 85% of each of the 'new battery' life spans gives the following values:

	Min.	Q$_1$	Median	Q$_3$	Max.
New batteries	19	23.5	26	28	31
85% of new life	16.1	20	22.1	23.8	26.4
5 year old batteries	18	23	24	26	30

All of the '5 year old' statistical measures are longer than 85% of the 'new battery' life spans, so the evidence supports the manufacturer's claim.

29 a The percentage of practices lasting between 3 and 4 hours is
$$\frac{6}{1+2+4+6+9+5+1} \times 100\%$$
$$= \frac{6}{28} \times 100\%$$
$$= 21.4\%$$

b

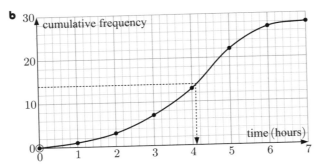

c The median ≈ 4.1 hours.

30 a The mid-interval value $= 85$ cm.

b **i** Using mid-interval values as an estimate for each class, the mean height $\overline{h} \approx 109$ cm.

    ii $s \approx 12.1$ cm

c The value 3 standard deviations above the mean
$$\approx 109 + 3 \times 12.1$$
$$\approx 145 \text{ cm}$$

31 a The data is discrete.

c **i** The median $= 2$ children.

    ii The mean ≈ 2.17 children.

b

x	f
1	8
2	9
3	3
4	3
5	1

32 a The modal class is $3 \leqslant t \leqslant 4$ minutes.

b **i** The total number of calls made
$$= 2+6+8+13+11+8+5+4+2+2 = 61$$

    ii The average calls per month $= \frac{61}{3} \approx 20.3$

c $P(\text{call} \geqslant 6 \text{ minutes}) = \dfrac{5+4+2+2}{61} = \dfrac{13}{61} \approx 0.213$

33 a The data is continuous.

b Adding the frequencies, the total number of trees planted is 70.

c
$$\overline{h}$$
$$\approx \frac{5.5\times2+6.5\times4+7.5\times7+8.5\times13+9.5\times12+10.5\times20+11.5\times6+12.5\times4+13.5\times2}{70}$$
$$\approx \frac{670}{70}$$
$$\approx 9.57 \text{ cm}$$

34 a 36 competitors ran the race.

b The median corresponds to $\text{CF} = \frac{36}{2} = 18$

    $\therefore$ the median ≈ 58

c Q$_1$ is the 25th percentile.
When the cumulative frequency $= 0.25 \times 36 = 9$, $t \approx 51$.

    $\therefore$ Q$_1 \approx 51$

Q$_3$ is the 75th percentile.
When the cumulative frequency $= 0.75 \times 36 = 27$, $t \approx 67$.

    $\therefore$ Q$_3 \approx 67$

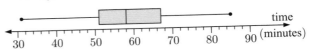

35 a Using technology:

    i $\overline{x} \approx 5.47$ **ii** $s \approx 1.39$

b **i** The percentage of scores which are 5 or less
$$= \frac{5+7}{19} \times 100\%$$
$$= \frac{12}{19} \times 100\%$$
$$\approx 63.2\%$$

    ii The percentage of scores greater than 5
$$\approx 100\% - 63.2\%$$
$$\approx 36.8\%$$

36 a The variable is discrete.

b The total number of goals $= 15 \times 2.4 = 36$

c Let x be the number of goals scored in the final game.
$$\therefore \quad \frac{36+x}{16} = 3$$
$$\therefore \quad 36+x = 48$$
$$\therefore \quad x = 12$$

Hence, 12 goals are needed for the average to be 3.0.

LONG QUESTIONS

1 a

No. of weeds	Frequency	Cumulative frequency
0 - 4	9	9
5 - 9	15	24
10 - 14	10	34
15 - 19	9	43
20 - 24	5	48
25 - 29	2	50

b **i** $P(\text{no more than } 19) = \frac{43}{50}$

    ii $P(9 \text{ or fewer}) = \frac{24}{50}$

c

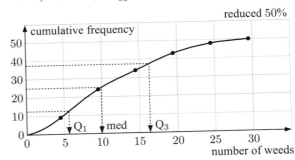

d From the graph:

    i median ≈ 10 **ii** first quartile ≈ 6

    iii 75th percentile ≈ 17

e interquartile range $\approx 17 - 6 \approx 11$

f mean is ≈ 11.2 {using technology}

g
$$12\,000 \text{ m}^2 = 120\,000\,000 \text{ cm}^2$$
$$\therefore \quad \text{estimated number of weeds} = \left(\frac{120\,000\,000}{80}\right) \times 11.2$$
$$= 16\,800\,000$$

2 a $\dfrac{n+1}{2} = \dfrac{25+1}{2} = 13$

So, the median is the 13th value. $\therefore$ median $= \$293$
Range $= 316 - 271 = 45$
If we leave out the median, then Q_1 is in the middle of the lower 12 values (between the 6th and 7th values), and Q_3 is in the middle of the higher 12 values (between the 19th and 20th values).
So, $Q_1 = \$283$ and $Q_3 = \$301$.

b
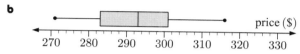
price ($)

c
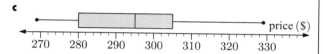
price ($)

d The new prices are more spread out.

e The mean price has increased slightly. Some printers have increased in price, others have decreased.

f The mean price has increased by \$2 from \$293 to \$295.
Percentage increase is $\frac{2}{293} \times 100\% \approx 0.683\%$.

3 a Mean $= \dfrac{3.6 + 4.6 + 5.6 + ... + 6.1 + 6.4}{40} = 5.265$

Using technology, the standard deviation is 0.856.

b

Distance (m)	Cumulative frequency
$3.5 \leqslant d < 4.0$	4
$4.0 \leqslant d < 4.5$	7
$4.5 \leqslant d < 5.0$	15
$5.0 \leqslant d < 5.5$	22
$5.5 \leqslant d < 6.0$	31
$6.0 \leqslant d < 6.5$	39
$6.5 \leqslant d < 7.0$	39
$7.0 \leqslant d < 7.5$	40

c
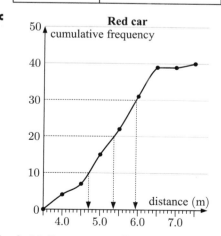
Red car

d i Median ≈ 5.4 **ii** $Q_1 \approx 4.6$ **iii** $Q_3 \approx 5.9$

e
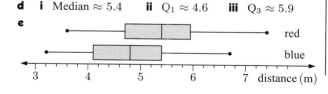
red
blue
distance (m)

f All values of the five-number summary (min, Q_1, median, Q_3, and max) for the red car are higher than those for the blue car. This evidence is strongly against the view that the cars were made by the same machine.

4 a Mean $= \dfrac{312 + 320 + 326 + ... + 321 + 314 + 324}{24}$
≈ 319 g
Range $= 334 - 306 = 28$ g

b

Weight (g)	Frequency		Weight (g)	Frequency
305 - 309	5		320 - 324	5
310 - 314	3		325 - 329	4
315 - 319	3		330 - 334	4

c
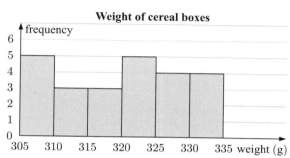
Weight of cereal boxes

d The stated weight is higher than the average (mean) contents, but the sample size is insufficient to make a strong conclusion.

e i Mean
$= \dfrac{312 \times 3 + 317 \times 5 + 322 \times 8 + 327 \times 6 + 332 \times 2}{24}$
≈ 322 g

ii The mean weight is now higher and the spread of weights is less. The evidence is in favour of an improvement in the production process.

5 a
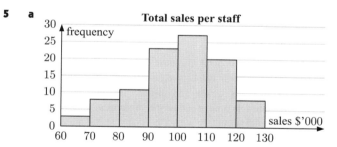
Total sales per staff

b minimum is \$120\,000, maximum is \$129\,999.99

c Mean $= \dfrac{65 \times 3 + 75 \times 8 + ... + 125 \times 8}{100}$ \$'000
$= \$100\,500$
Using technology, $s \approx \$14\,800$.

d i $\overline{x} + 2s \approx 100\,500 + 2 \times 14\,800 \approx 130\,100$
A minimum of \$130\,100 in sales was needed.

ii No bonuses were paid.

e The top 8 sales staff sold more than \$120\,000.
This is $\$120\,000 - \$100\,500 = \$19\,500$ above the mean and $\frac{19\,500}{14\,800} \approx 1.32$ standard deviations above the mean.

f $\overline{x} - 1.385s \approx 100\,500 - 1.385 \times 14\,800 \approx \$80\,002$
So, 11 staff will lose their jobs.

6 a modal class is $180 \leqslant t < 240$ seconds

b $x = 25 + 40 = 65, \quad y = 65 + 60 = 125$

c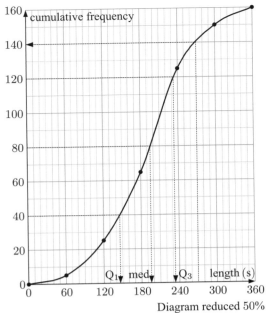

Diagram reduced 50%

d   **i** median ≈ 195 s
    ii $Q_1 \approx 147$ s, $Q_3 = 232$ s,
       so IQR $= Q_3 - Q_1 \approx 85$ s

e 4.5 min $= 270$ s
    From graph, CF ≈ 140
    So, P(song no longer than 4.5 min) $\approx \frac{140}{160} \approx \frac{7}{8}$

f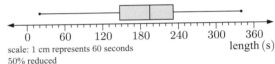

scale: 1 cm represents 60 seconds
50% reduced

7 **a**   **i** $Q_1 = 4$ **ii** median $= 6$ **iii** $Q_3 = 12.5$

   b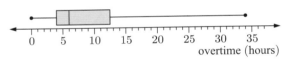

c The data is not symmetric, but rather is positively skewed. The median value is 2 hours above the first quartile, but 6.5 hours below the upper quartile. Additionally, the median is 6 hours above the minimum, but 28 hours below the maximum.

d   **i** Range $= 34 - 0$ **ii** IQR $= Q_3 - Q_1$
             $= 34$ hours $= 12.5 - 4$
                                         $= 8.5$ hours

e   **i** *"The central 50% of employees work between 4 and 12.5 hours of overtime."*

    ii *"75% of employees work no more than 12.5 hours overtime."*

f Median $= 6$ hours   {from part **a**}
    Mode $= 5$ hours   Mean $= 9.36$ hours
    The law firm would quote the smallest measure of 'centre' in order to downplay the amount of overtime worked. They would choose the median, as it is only 5 hours.

8 **a** 90 games were played.

   b The median corresponds to CF $= 45$.
      Hence, the median ≈ 85 minutes.

---

c IQR $= Q_3 - Q_1$
    Q_3 corresponds to CF $= 67.5$ which is 110 minutes.
    Q_1 corresponds to CF $= 22.5$ which is 66 minutes.
    Using the CF curve, IQR $\approx 110 - 66 = 44$ minutes.

d The 10th percentile corresponds to CF $= 9$ which is 49 minutes.

e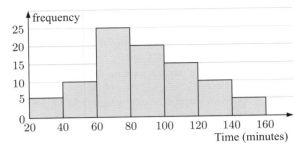

f   **i** mean ≈ 30.5 minutes
    ii standard deviation ≈ 30.5 minutes

9 **a** 170 feeds

   b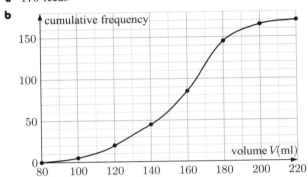

c   **i** The median corresponds to the 85th value, which is 160 mL.
    ii IQR $= Q_3 - Q_1 \approx 174 - 138 \approx 36$ mL

d Using the mid-interval values and frequency values:
    i $\overline{v} \approx 155$ mL   **ii** $s \approx 26.7$ mL

e   **i** $p \approx 128$ mL, $q \approx 182$ mL
    ii Number of feeds between 182 mL and 128 mL
        $= 148 - 29 \approx 119$
        $\therefore$ P$(p < v < q) \approx \frac{119}{170} \approx 0.7$

SOLUTIONS TO TOPIC 3 (SETS, LOGIC, AND PROBABILITY)

SHORT QUESTIONS

1 $A = \{0, 3, 6, 9\}$, $B = \{1, 2, 5, 10\}$
   a $A \cap B = \varnothing$
   b $A \cup B = \{0, 1, 2, 3, 5, 6, 9, 10\}$
   c $(A \cup B)' = \{4, 7, 8, 11\}$

2 **a** $A = \{4, 8, 12, 16\}$, $B = \{2, 4, 6, 8\}$
       $A \cup B = \{2, 4, 6, 8, 12, 16\}$
   b $A \cap B = \{4, 8\}$
   c $C = \{4, 8, 12\}$ (other answers are possible)

3 **a**   **i** -6 (or any integer)
      ii $3\frac{7}{10}$ (or any real number not an integer)
     iii 29 (or any non-negative integer)
     iv $\mathbb{Q}' \cap \mathbb{Z} = \varnothing$ so there are no elements in $\mathbb{Q}' \cap \mathbb{Z}$.

b

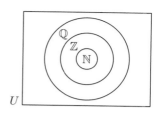

4 $A = \{9, 12, 15, 18\}, \quad B = \{9, 12, 18\}$

  a $A \cap B = \{9, 12, 18\}$

  b $A' = \{8, 10, 11, 13, 14, 16, 17\}$

  c

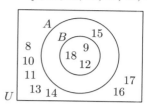

  d **i** $B \subset A$ is true

     ii $A' \cap B' = \{8, 10, 11, 13, 14, 16, 17\}$

       so $n(A' \cap B') = 7$ is true.

     iii Since $B \subset A$, $A \cup B = A$ is true.

5 $F = \{1, 2, 3, 4, 6, 8, 12\}$

  $P = \{2, 3, 5, 7, 11, 13, 17, 19\}$

  a $n(F) = 7$

  b $P \cap F = \{2, 3\}$

  c $P \cup F = \{1, 2, 3, 4, 5, 6, 7, 8, 11, 12, 13, 17, 19\}$

  d $P' \cap F = \{1, 4, 6, 8, 12\}$

6 **a**

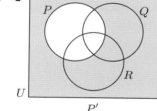

        P' $P \cup R$

  c **d**

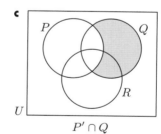

       $P' \cap Q$ $P \cap (Q \cup R)$

7 **a** **i** $25 + y + z$ students   **ii** x students

     iii $13 + x + y + z$ students

  b **i** P(student likes none of the foods)

$$= \frac{5}{61 + x + y + z}$$

     ii P(student likes rice | student likes noodles)

$$= \frac{13 + x}{36 + x + z}$$

8 **a** $U = \{1, 2, 3, 4, 5, 6, 7, 8, 9\}, \quad P = \{2, 4, 6, 8\},$

    $Q = \{2, 3, 5, 7\}$

    i $P \cap Q = \{2\}$ **ii** $(P \cup Q)' = \{1, 9\}$

b

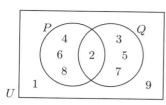

9 **a** $\neg r \wedge q \Rightarrow \neg p$

  b The converse of $q \vee \neg p \Rightarrow \neg r$ is $\neg r \Rightarrow q \vee \neg p$

    If Toby is not a dog then Toby has long ears or does not have four legs.

10 $P = \{1, 3, 5, 7, 9, 11\}, \quad Q = \{1, 2, 3, 4, 6, 12\},$

  $R = \{5, 10\}$

  a $P \cap Q = \{1, 3\}$ **b** $(P \cap Q) \cup R = \{1, 3, 5, 10\}$

  c $Q \cap R = \{\ \}$ **d** $P' \cap (Q \cup R) = \{2, 4, 6, 10, 12\}$

11 **a** $n(P) = 22$ **b** $n(P \cup Q) = 47$

  c $n(Q \cap R) = 13$ **d** $n((P \cup Q \cup R)') = 3$

  e $n(Q') = 30$ **f** $n(P \cap R) = 0$

12 **a** If E is the set of students studying English and S is the set of students studying Spanish then

$$n(E \cup S) = n(E) + n(S) - n(E \cap S)$$
$$= 18 + 25 - 6$$
$$= 37$$

    So, 37 IB students study English or Spanish.

  b Let F be the set of students taking lessons in French and E be those taking lessons in English. If all students must study either in French or English, then 100% study in one of these languages.

    $n(F \cup E) = n(F) + n(E) - n(E \cap F)$

    If $x\%$ is the percentage of students studying in both languages, then $100\% = 60\% + 76\% - x\%$

    $\therefore \ x\% = (60 + 76 - 100)\% = 36\%$

    So, 36% of students study in both languages.

  c 20% of 25 = 5 students do not study art or drama.

    Let A be the set of students studying art and D be those studying drama. This information is summarised in the Venn diagram, where x is the number studying both art and drama.

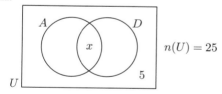

        $n(A \cup D) = 25 - 5 = 20$

      Now $n(A \cup D) = n(A) + n(D) - n(A \cap D)$

         $\therefore \ 20 = 13 + 9 - x$

         $\therefore \ x = 13 + 9 - 20 = 2$

    So, 2 students study both art and drama.

13 **a** **i** $\neg q \Rightarrow \neg r$

     ii $\neg p \Rightarrow \neg r \wedge \neg q$ or $\neg p \Rightarrow \neg(r \vee q)$

  b $\neg r \Rightarrow \neg(p \vee q)$: If Peter is not a student then he neither has black hair nor plays basketball.

14 **a** $n(P) = 4 + 1 + 2 + 4 = 11$

  b $n(M \cup D) - n(M \cap D) = 8 + 1 + 2 + 6 = 17$

  c $n(P \cap M') = 4 + 2 = 6$

  d $n(D') = 8 + 1 + 4 + 3 = 16$

15 $A = \{2, 3, 4, 6, 8, 12, 16, 24, 48\}$
$B = \{6, 12, 18, 24, 30, 36, 42, 48\}$
$C = \{8, 16, 24, 32, 40, 48\}$

  a $A \cap B \cap C = \{24, 48\}$ **b** $n(B) = 8$

  c $A' \cap B = \{18, 30, 36, 42\}$

16 **a** **i** $Y = \{3, 4, 5, 6, 7\}$ **ii** $Z = \{0, 2, 4, 6, 8\}$

  b

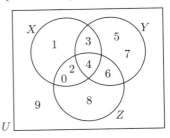

17 **a** **i** $\neg p \wedge \neg q$: Wilson is not an active dog and Wilson does not dig holes.

    ii $p \veebar q$: Wilson is an active dog or Wilson digs holes but not both.

    iii $\neg p \Rightarrow q$: If Wilson is not an active dog then Wilson digs holes.

  b

p	q	$\neg p$	$\neg p \Rightarrow q$
T	T	F	T
T	F	F	T
F	T	T	T
F	F	T	F

$\neg p \Rightarrow q$ is neither a contradiction nor a tautology.

18 **a** Since b is a prime number less than 10, $b = 2$.

  b $\mathbb{Q}$ is the set of even numbers less than 10.

  c $a = 9$

19 **a** Converse $q \Rightarrow p$: If I live near the sea then I like the beach.

  b Inverse $\neg p \Rightarrow \neg q$: If I do not like the beach then I do not live near the sea.

  c Contrapositive $\neg q \Rightarrow \neg p$: If I do not live near the sea then I do not like the beach.

20 **a** **i** $A \cap B = \varnothing$ is false as A and B have common elements.

    ii $A \cap C = C$ is true as C lies entirely within A.

    iii $B \subset A'$ is false as A and B have common elements.

    iv $C \subset (A \cap B)$ is false as C has no elements in common with $A \cap B$.

  b The shaded region is $A' \cap B$.

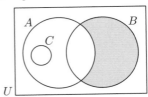

21 **a** $n(S \cap B \cap G) = 8$ members play all three sports.

  b $n(S \cap B') + n(S' \cap B)$
$= (30 + 12) + (20 + 9)$
$= 71$ play soccer or basketball but not both.

  c $n(G \cap S' \cap B') = 14$ play only golf.

  d $n(B') = 30 + 12 + 14$
$= 56$ do not play basketball.

  e $n(\text{play two or three sports}) = 12 + 6 + 9 + 8$
$= 35$ members

f $n(S \cap G') = 30 + 6 = 36$ play soccer but not golf.

22 **a** **i** $p \Rightarrow q$: If the sun is shining then I take the dog for a walk.

    ii $\neg p \vee q$: The sun is not shining or I take the dog for a walk.

  b The converse of $p \Rightarrow q$ is $q \Rightarrow p$: If I take the dog for a walk then the sun is shining.

  c $\neg q \Rightarrow \neg p$

  d $\neg q \Rightarrow \neg p$ is the contrapositive of $p \Rightarrow q$.

23 **a** $\neg p \Rightarrow \neg q$: If Molly does not have a DVD player then Molly does not like watching movies.

  b

p	q	$q \Rightarrow p$
T	T	T
T	F	T
F	T	F
F	F	T

  c

p	q	$\neg p$	$\neg q$	$\neg p \Rightarrow \neg q$
T	T	F	F	T
T	F	F	T	T
F	T	T	F	F
F	F	T	T	T

  d The final column is the same for both propositions, so $q \Rightarrow p$ and $\neg p \Rightarrow \neg q$ are logically equivalent.

24 **a**

p	q	$p \wedge q$	$(p \wedge q) \Rightarrow p$
T	T	T	T
T	F	F	T
F	T	F	T
F	F	F	T

  b The final column only has T values, so this is a tautology.

25 P(a randomly selected packet contains more than 40 biscuits)
$= \frac{24}{500} = \frac{6}{125}$

  a P(both contain more than 40 biscuits)
$= \frac{6}{125} \times \frac{6}{125} \approx 0.002\,30$

  b P(one contains more than 40 biscuits and the other less than or equal to 40)
$= 2 \times \frac{6}{125} \times \frac{119}{125} \approx 0.0914$

26

p	q	$\neg q$	$p \Rightarrow \neg q$	$\neg p$	$\neg p \Rightarrow q$	$(p \Rightarrow \neg q) \vee (\neg p \Rightarrow q)$
T	T	F	F	F	T	T
T	F	T	T	F	T	T
F	T	F	T	T	T	T
F	F	T	T	T	F	T

This is a tautology.

27 **a** **i** $(\neg r \wedge \neg q) \vee p$: I do not buy an ice cream and I have no money or the weather is hot.

    ii $(p \wedge q) \Rightarrow r$: If the weather is hot and I have money then I buy an ice cream.

    iii $(\neg p \vee \neg q) \Rightarrow \neg r$: If the weather is not hot or I have no money then I do not buy an ice cream.

  b The contrapositive of $p \Rightarrow r$ is $\neg r \Rightarrow \neg p$: If I do not buy an ice cream then the weather is not hot.

28 **a** $U = \{0, 1, 2, 3, 4, 5, 6, 7, 8, 9\}$,
$P = \{2, 3, 5, 7\}$, $Q = \{0, 3, 6, 9\}$

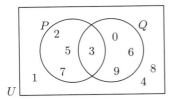

b **i** $P(Q) = \frac{4}{10} = \frac{2}{5}$ **ii** $P(P \cap Q') = \frac{3}{10}$

iii $P(Q \mid P) = \frac{1}{4}$

29 **a** $\neg q$: I do not visit the doctor.

b **i** $\neg p \Rightarrow q$: If I do not eat an apple every day then I visit the doctor.

ii $(\neg p \vee q) \Rightarrow q$: If I do not eat an apple every day or I visit the doctor then I visit the doctor.

c

p	q	$\neg p$	$\neg p \vee q$	$(\neg p \vee q) \Rightarrow q$
T	T	F	T	T
T	F	F	F	T
F	T	T	T	T
F	F	T	T	F

As the last column contains some F values and some T values, this is neither a tautology nor a contradiction.

30 **a**

a	b	$a \Leftrightarrow b$	$(a \Leftrightarrow b) \wedge b$	$(a \Leftrightarrow b) \wedge b \Rightarrow a$
T	T	T	T	T
T	F	F	F	T
F	T	F	F	T
F	F	T	F	T

b Since the last column contains all T values, this is a tautology.

31 **a**

p	q	$p \wedge q$	$\neg(p \wedge q)$	$\neg(p \wedge q) \wedge q$	$\neg p \vee q$	$\neg(p \wedge q) \wedge q \Rightarrow (\neg p \vee q)$
T	T	T	F	F	T	T
T	F	F	T	F	F	T
F	T	F	T	T	T	T
F	F	F	T	F	T	T

b Since the last column contains all T values, this is a tautology.

32 **a**

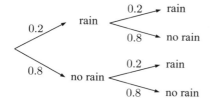

b We assume that the probability of it raining each day is independent.

c **i** P(raining two days in a row) $= 0.2 \times 0.2$

$= 0.04$

ii P(raining on one day only)

$= \text{P}((\text{R and NR}) \text{ or } (\text{NR and R}))$

$= \text{P}(\text{R and NR}) + \text{P}(\text{NR and R})$

$= 0.2 \times 0.8 + 0.8 \times 0.2$

$= 0.16 + 0.16$

$= 0.32$

33 **a** **i** $A = \{2, 3, 5, 6\}$ and $n(A) = 4$

ii $A \cap B = \{3, 6\}$ and $n(A \cap B) = 2$

b **i** $A \cup B \cup C = \{2, 3, 4, 5, 6, 7, 8\}$

ii $(A' \cap B) \cup C = \{4, 5, 6, 7, 8\}$

34 **a** **i** $p \Rightarrow q$: If the sink is blocked then the plumber will clear the drain.

ii $\neg q \Rightarrow \neg p$: If the plumber will not clear the drain then the sink is not blocked.

b

p	q	$p \Rightarrow q$	$\neg q \Rightarrow \neg p$
T	T	T	T
T	F	F	F
F	T	T	T
F	F	T	T

The truth columns are the same, so the statements are logically equivalent.

35 **a**

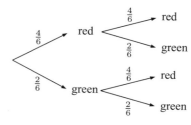

b **i** P(two reds) $= \frac{2}{3} \times \frac{2}{3} = \frac{4}{9}$

ii P(two greens) $= \frac{1}{3} \times \frac{1}{3} = \frac{1}{9}$

iii P(one of each) $= \frac{2}{3} \times \frac{1}{3} + \frac{1}{3} \times \frac{2}{3} = \frac{4}{9}$

36 **a** **i** $X = \{a, c, e, h, i, m, s, t\}$

ii $Y = \{c, e, g, i, l, n, t\}$

b

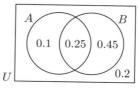

c **i** $P(X \cup Y) = \frac{11}{26}$ **ii** $P(X \cap Y') = \frac{4}{26} = \frac{2}{13}$

37 **a** $P(A \cup B) = P(A) + P(B) - P(A \cap B)$

$\therefore \quad 0.8 = 0.35 + 0.7 - P(A \cap B)$

$\therefore \quad P(A \cap B) = 1.05 - 0.8 = 0.25$

b

A Venn diagram with regions labelled 0.1, 0.25, 0.45 and 0.2 outside in U.

c $P(A' \cap B') = 0.2$

d If A and B are independent then

$$P(A \cap B) = P(A) \times P(B)$$

$$P(A) \times P(B) = 0.35 \times 0.7$$

$$= 0.245$$

but $P(A \cap B) = 0.25$

$\therefore$ events A and B are not independent.

38 **a**

p	q	$\neg p$	$\neg q$	$p \Rightarrow q$	$(p \Rightarrow q) \wedge \neg q$	$(p \Rightarrow q) \wedge \neg q \Rightarrow \neg p$
T	T	F	F	T	F	T
T	F	F	T	F	F	T
F	T	T	F	T	F	T
F	F	T	T	T	T	T

b The statement is $(p \Rightarrow q) \wedge \neg q \Rightarrow \neg p$.

We know from **a** that this statement is a tautology.

So, the statement is valid.

39 **a** Number in class $= 7 + 9 + 3 + 4$

$= 23$

b $n(C \cap D) = 9$ **c** $n(D) - n(D \cap C) = 7$

d $n(C \cup D) = 7 + 9 + 3 = 19$

e $n((C \cup D)') = 4$

40 a

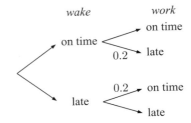

wake   *work*

b i P(Katie will be late for work tomorrow)
= P(awake on time ∩ late to work)
    + P(wakes up late ∩ late to work)
= $(0.7 \times 0.2) + (0.3 \times 0.8) = 0.38$

  ii P(Katie will be on time to work tomorrow)
= 1 − P(Katie will be late to work tomorrow)
= $1 - 0.38 = 0.62$

41 a $c = 1 - (0.3 + 0.4 + 0.2) = 0.1$

b i $P(A \cup B) = 0.3 + 0.4 + 0.2 = 0.9$

  ii $P(A \cap B) = 0.4$ **iii** $P(A' \cap B) = 0.2$

  iv $P(A \mid B) = \frac{0.4}{0.6} = \frac{2}{3}$ **v** $P(B \mid A) = \frac{0.4}{0.7} = \frac{4}{7}$

42 a i $a \Rightarrow b$: If the lights are on then somebody is at home.

  ii $\neg b \wedge \neg a$: Nobody is at home and the lights are not on.

  iii $a \veebar b$: The lights are on or somebody is at home, but not both.

b i $\neg a \Rightarrow \neg b$: If the lights are not on then nobody is at home.

  ii $\neg b \Rightarrow \neg a$: If nobody is at home then the lights are not on.

c

a	b	$\neg a$	$\neg b$	$\neg a \Rightarrow \neg b$	$b \Rightarrow a$
T	T	F	F	T	T
T	F	F	T	T	T
F	T	T	F	F	F
F	F	T	T	T	T

The truth columns for the inverse $\neg a \Rightarrow \neg b$ and converse $b \Rightarrow a$ are the same, so these statements are logically equivalent.

43

	soft	medium	hard	*Total*
dark	10	24	15	49
light	18	14	9	41
Total	28	38	24	90

a i P(dark or hard) $= \frac{58}{90} = \frac{29}{45} \approx 0.644$

  ii P(light | soft) $= \frac{18}{28} = \frac{9}{14} \approx 0.643$

b i P(both medium) $= \frac{38}{90} \times \frac{37}{89} \approx 0.176$

  ii P(at least one is soft and light)
= 1 − P(neither is soft and light)
$= 1 - \frac{72}{90} \times \frac{71}{89}$
≈ 0.362

44 a

p	q	$p \wedge q$	$p \veebar q$	$(p \wedge q) \vee (p \veebar q)$
T	T	T	F	T
T	F	F	T	T
F	T	F	T	T
F	F	F	F	F

b i $x^2 - 16 = 0$ is true if $x = \pm 4$

  ii $p \wedge q$ is true if $x \geqslant 1$ *and* $x^2 - 16 = 0$
So, $x = 4$.

  iii Using the truth table in **a**, we see that
$(p \wedge q) \vee (p \veebar q)$ is logically equivalent to $p \vee q$.
∴ $x \geqslant 1$ or $x^2 - 16 = 0$ or both
∴ $x \geqslant 1$ or $x = -4$.

45 a $n(A \cap Z) = 4$ **b** $n(A \cup B) = 11 + 8 = 19$

c $n((B \cap W) \cup (C \cap Z)) = 3 + 2 = 5$

d $P(A \cap B) = 0$

e $P((C \cap W) \mid (B \cup C)) = \dfrac{11}{3 + 5 + 11 + 2}$
$= \frac{11}{21} \approx 0.524$

46 a p: Kania cooks the meal.
    q: The meal is delicious.
    r: Mal is happy.

b $(p \Rightarrow q) \wedge (q \Rightarrow r) \wedge r \Rightarrow p$

c

p	q	r	$p \Rightarrow q$	$q \Rightarrow r$	$(p \Rightarrow q) \wedge (q \Rightarrow r) \wedge r$	$(p \Rightarrow q) \wedge (q \Rightarrow r) \wedge r \Rightarrow p$
T	T	T	T	T	T	T
T	T	F	T	F	F	T
T	F	T	F	T	F	T
T	F	F	F	T	F	T
F	T	T	T	T	T	F
F	T	F	T	F	F	T
F	F	T	T	T	F	T
F	F	F	T	T	F	T

The argument is not a tautology.

LONG QUESTIONS

1 a i $P = \{2, 3, 5, 7, 11, 13\}$ **ii** $Q = \{5, 10, 15\}$

  iii $R = \{1, 3, 5, 15\}$

b i $P \cap Q \cap R = \{5\}$

  ii $(P \cup Q \cup R)' = \{4, 6, 8, 9, 12, 14\}$

c

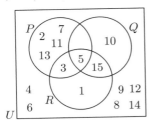

d i $P(Q) = \frac{3}{15} = \frac{1}{5}$ **ii** $P(R') = \frac{11}{15}$

  iii $P(Q \cup R) = \frac{5}{15} = \frac{1}{3}$ **iv** $P(P \cap (Q \cup R)') = \frac{4}{15}$

e The only number which is both prime and a multiple of 5 is 5 itself, but 5 is also a factor of 15, so no element of U satisfies these conditions.

2 a i $P = \{0, 2, 4, 6, 8, 10\}$ **ii** $Q = \{0, 3, 6, 9\}$

  iii $R = \{1, 2, 3, 4, 6\}$ **iv** $P \cap Q \cap R = \{6\}$

b

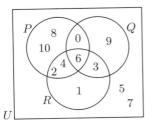

c i $P \cup Q$: The elements that are multiples of 2 or 3 or both.

ii $P' \cap Q' \cap R'$: The elements that are neither multiples of 2 nor of 3 nor factors of 12.

d i

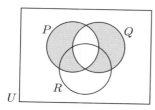

ii

p	q	r	$p \wedge r$	$p \veebar q$	$(p \wedge r) \Rightarrow (p \veebar q)$
T	T	T	T	F	F
T	T	F	F	F	T
T	F	T	T	T	T
T	F	F	F	T	T
F	T	T	F	T	T
F	T	F	F	T	T
F	F	T	F	F	T
F	F	F	F	F	T

iii An element of U for which $(p \wedge r) \Rightarrow (p \veebar q)$ is true is 2. This corresponds to the third entry in the table.

3 a i $p \veebar q$

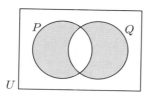

ii $\neg(p \wedge q)$

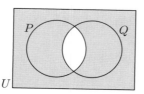

iii $\neg p \vee \neg q$

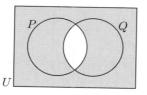

b Since **a ii** and **a iii** have the same areas shaded, $\neg(p \wedge q)$ and $\neg p \vee \neg q$ are logically equivalent.

c $\neg(p \wedge q) \Rightarrow \neg q$: If the bush is not a thorny rose bush then it is not a rose bush.

d

p	q	$p \wedge q$	$\neg(p \wedge q)$	$\neg q$	$\neg(p \wedge q) \Rightarrow \neg q$
T	T	T	F	F	T
T	F	F	T	T	T
F	T	F	T	F	F
F	F	F	T	T	T

The result is neither a tautology nor a contradiction.

4 a

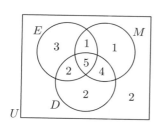

b i 6 critics enjoyed only one course.

ii 12 critics enjoyed two or more courses.

iii 2 critics enjoyed none of the courses.

c i P$(M) = \frac{11}{20}$

ii P$(M \cap E) = \frac{6}{20} = \frac{3}{10}$

iii P$(E \cup D) = \frac{17}{20}$

d i P(both enjoyed dessert) $= \frac{13}{20} \times \frac{12}{19} \approx 0.41053$
$$\approx 0.411$$

ii P(at least one enjoyed dessert)
$= 1 - $ P(neither enjoyed dessert)
$= 1 - \frac{7}{20} \times \frac{6}{19}$
≈ 0.88947
≈ 0.889

iii P(both enjoyed dessert | at least one enjoyed dessert)
$$= \frac{\text{P(both enjoyed dessert)}}{\text{P(at least one enjoyed dessert)}}$$
$\approx \frac{0.41053}{0.88947}$ {**d i** and **d ii**}
≈ 0.462

5 a i $r \Rightarrow p$ **ii** $q \wedge p \Rightarrow r$

b i The number must be an odd prime greater than 2. So, the truth set is $\{3, 5, 7\}$.

ii The number is either odd or a prime greater than 2, but not both. So, the truth set is $\{1, 9\}$.

iii The number is an even multiple of 3, or a prime greater than 2, or both.
So, the truth set is $\{3, 5, 6, 7\}$.

c i If x is an odd number and x is not a multiple of 3, then x is not a prime number greater than 2.

ii

p	q	r	$p \wedge \neg q$	$\neg r$	$p \wedge \neg q \Rightarrow \neg r$
T	T	T	F	F	T
T	T	F	F	T	T
T	F	T	T	F	F
T	F	F	T	T	T
F	T	T	F	F	T
F	T	F	F	T	T
F	F	T	F	F	T
F	F	F	F	T	T

iii The statement is false when p and r are true and q is false.

iv The number must be odd and a prime number greater than 2, but it is not a multiple of 3. So, $x = 5$ or 7.

6 a Let A be the students enrolled in Art,
B be the students enrolled in Biology,
and E be the students enrolled in Economics.
Let the number who study all three subjects be x. We use the information given to begin constructing a Venn diagram.

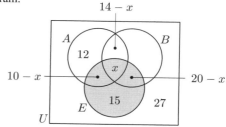

Now $n(E) = 38$,

$$15 + (20 - x) + (10 - x) + x = 38$$
$$\therefore \quad -x + 45 = 38$$
$$\therefore \quad x = 7$$

So, 7 students study all 3 subjects.

b $n(\text{study only Biology})$
$= 92 - (27 + 38 + (14 - 7) + 12)$
$= 8$

c

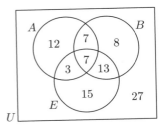

d **i** $P(E) = \frac{38}{92} = \frac{19}{46}$

  ii $P(A \mid \text{study at least one subject}) = \frac{29}{92-27} = \frac{29}{65}$

e **i** $P(\text{both study biology}) = \frac{35}{92} \times \frac{34}{91} \approx 0.142$

  ii $P(\text{both study exactly two subjects})$
  $= \frac{23}{92} \times \frac{22}{91} \approx 0.0604$

7 $\neg(p \lor q) \Rightarrow \neg p \land \neg q$

a **i**

p	q	$\neg p$	$\neg q$	$p \lor q$	$\neg(p \lor q)$	$\neg p \land \neg q$	$\neg(p \lor q) \Rightarrow \neg p \land \neg q$
T	T	F	F	T	F	F	T
T	F	F	T	T	F	F	T
F	T	T	F	T	F	F	T
F	F	T	T	F	T	T	T

  ii This is a tautology.

b **i** $p \land q$: I work hard and I get a promotion.

  ii $\neg p \Rightarrow \neg q$:
  If I do not work hard then I do not get a promotion.

c **i**

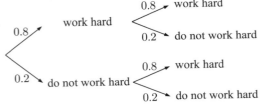

  ii **(1)** $P(\text{both work hard}) = 0.8 \times 0.8$
  $= 0.64$

  (2) $P(\text{only one works hard})$
  $= 0.8 \times 0.2 + 0.2 \times 0.8$
  $= 0.16 + 0.16$
  $= 0.32$

8 **a**

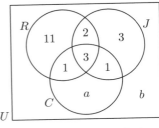

  i Since 7 students like Classical music,
  $a = 7 - (1 + 3 + 1) = 2$

---

  ii The region containing 11 students are those who only like Rock music.

  iii Since there are 30 students in total,
  $n(b) = 30 - (11 + 2 + 3 + 1 + 3 + 1 + 2)$
  $= 30 - 23$
  $= 7$

  iv Region b consists of the students who do not like any of these music genres.

b $R' \cap J$

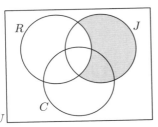

c **i** $P(\text{likes all 3 types}) = \frac{3}{30} = \frac{1}{10}$

  ii $P(\text{likes only Classical music}) = \frac{2}{30} = \frac{1}{15}$

  iii $P(R \mid J) = \frac{5}{9}$

d **i** $P(\text{both like Rock music only}) = \frac{11}{30} \times \frac{10}{29} \approx 0.126$

  ii $P(\text{one likes Rock only}$
  $\text{and the other likes all 3 types})$
  $= \left(\frac{11}{30} \times \frac{3}{29}\right) + \left(\frac{3}{30} \times \frac{11}{29}\right)$
  ≈ 0.0759

9 **a** **i** If I play computer games then I go to sleep.

  ii If I do not play computer games then I do not go to sleep.

b inverse

c $p \Rightarrow (q \land r)$

d

p	q	r	$q \land r$	$p \Rightarrow (q \land r)$	$(p \Rightarrow (q \land r)) \land p$	$\begin{array}{c}(p \Rightarrow (q \land r))\\ \land p \Rightarrow r\end{array}$
T	T	T	T	T	T	T
T	T	F	F	F	F	T
T	F	T	F	F	F	T
T	F	F	F	F	F	T
F	T	T	T	T	F	T
F	T	F	F	T	F	T
F	F	T	F	T	F	T
F	F	F	F	T	F	T

e Since the last column contains all Ts, this is a tautology.

10 **a** **i** There were $11 + 25 + 34 + 30 + 20 = 120$ students in the survey.

  ii $P(\text{spends less than \$20 per week}) = \frac{36}{120} = \frac{3}{10}$

  iii $P(\text{spends between \$30 and \$50 per week}) = \frac{50}{120}$
  $= \frac{5}{12}$

  $\therefore$ we expect $1500 \times \frac{5}{12} = 625$ students to spend between \$30 and \$50 per week.

b $P(\text{spends more than \$30} \mid \text{spends more than \$10})$
  $= \frac{50}{109} \approx 0.459$

c Let F be the outcome that a given student spends more than \$40 per week.

  i $P(FFF) = \frac{20}{120} \times \frac{19}{119} \times \frac{18}{118} \approx 0.004\,06$

  ii $P(F'F'F') = \frac{100}{120} \times \frac{99}{119} \times \frac{98}{118} \approx 0.575\,77$
  ≈ 0.576

iii P(exactly one spends more than $40)

$= P(FF'F') + P(F'FF') + P(F'F'F)$

$= \left(\frac{20}{120} \times \frac{100}{119} \times \frac{99}{118}\right) + \left(\frac{100}{120} \times \frac{20}{119} \times \frac{99}{118}\right)$

$\quad + \left(\frac{100}{120} \times \frac{99}{119} \times \frac{20}{118}\right)$

≈ 0.35251

≈ 0.353

iv P(at least two spend more than $40)

$= 1 - $ P(none spends more than $40)

$\quad - $ P(exactly one spends more than $40)

$\approx 1 - 0.57577 - 0.35251$ {using **ii** and **iii**}

≈ 0.0717

11 a

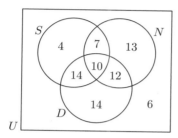

b $4 + 7 + 10 + 14 + 13 + 12 + 14 = 74$

$\therefore 80 - 74 = 6$ students watched none.

c **i** P(watched only Drama) $= \frac{14}{80} = \frac{7}{40}$

ii $P(S \mid N) = \frac{17}{42}$

d **i** $p \Rightarrow q$: If you watch sport on television then you like sport.

ii $q \Rightarrow p$: If you like sport then you watch sport on television.

iii $\neg q \Rightarrow \neg p$

iv $\neg q \Rightarrow \neg p$ is the contrapositive of $p \Rightarrow q$.

12 a

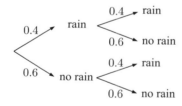

b **i** P(rain on both days) $= 0.4 \times 0.4$

$= 0.16$

ii P(no rain on one day) $= 0.4 \times 0.6 + 0.6 \times 0.4$

$= 0.24 + 0.24$

$= 0.48$

c P(fine on 5 consecutive days)

$= 0.6 \times 0.6 \times 0.6 \times 0.6 \times 0.6$

≈ 0.0778

d **i** $p \Rightarrow q$: If it is raining then I wear my raincoat.

ii $\neg q \Rightarrow \neg p$: If I do not wear my raincoat then it is not raining.

e **i**

p	q	$p \wedge q$	$\neg(p \wedge q)$	$\neg p$	$\neg(p \wedge q) \Rightarrow \neg p$
T	T	T	F	F	T
T	F	F	T	F	F
F	T	F	T	T	T
F	F	F	T	T	T

ii The proposition is false when it is raining and I do not wear my raincoat.

13 a There were $\quad 40 + 160 + 200 + 120 + 100 + 60 + 20$

$= 700$ patrons surveyed.

b **i** P(seated within 4 min) $= \frac{520}{700} \approx 0.743$

ii P(not seated within 1 min)

$= 1 - $ P(seated within 1 min)

$= 1 - \frac{40}{700}$

≈ 0.943

iii P(seated between 1 and 4 mins) $= \frac{480}{700} \approx 0.686$

c **i** The number of seats in each row follows the sequence 22, 26, 30, 34, ... which is arithmetic with first term 22 and common difference 4.

$\therefore \ u_n = 22 + (n-1)4$

$= 4n + 18$

ii $S_n = \frac{n}{2}(2u_1 + (n-1)d)$

$\therefore \ S_{40} = 20(2 \times 22 + 39 \times 4)$

$= 4000$

So, there are 4000 seats in the concert hall.

d **i** $u_{40} = 4 \times 40 + 18 = 178$

$\therefore$ P(seated in back row) $= \frac{178}{4000} = 0.0445$

ii $S_{10} = \frac{10}{2}(2 \times 22 + 9 \times 4)$

$= 400$

$\therefore$ P(seated in first 10 rows) $= \frac{400}{4000} = \frac{1}{10}$

14 a **i**

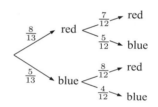

ii **(1)** P(two marbles same colour)

$= \left(\frac{8}{13} \times \frac{7}{12}\right) + \left(\frac{5}{13} \times \frac{4}{12}\right)$

$= \frac{76}{156}$

≈ 0.487

(2) P(at least one blue marble) $= 1 - $ P(no blue)

$= 1 - $ P(two red)

$= 1 - \frac{8}{13} \times \frac{7}{12}$

$= \frac{100}{156} \approx 0.641$

b **i**

ii **(1)** P(even number on the spinner) $= \frac{2}{5}$

(2) P(two blue marbles)

$= $ P(EBB or OBB)

$= $ P(EBB) + P(OBB)

$= \frac{2}{5} \times \frac{5}{13} \times \frac{4}{12} + \frac{3}{5} \times \frac{4}{11} \times \frac{3}{10}$

≈ 0.117

15 a i P(two chocolates) $= \frac{4}{11}$ **ii** $\frac{3}{10} + x = 1$

$\therefore\ x = \frac{7}{10}$

iii P(both hard) $= \frac{4}{11} \times \frac{3}{10}$

$= \frac{6}{55}$

iv P(one of each type) $= \frac{4}{11} \times \frac{7}{10}$

$= \frac{14}{55}$

b i

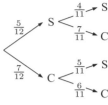

ii (1) P(both strawberry) $= \frac{5}{12} \times \frac{4}{11}$

$= \frac{20}{132}$

≈ 0.152

(2) P(second is strawberry) $= \frac{5}{12} \times \frac{4}{11} + \frac{7}{12} \times \frac{5}{11}$

≈ 0.417

iii The girl selects 3 cherry chocolates in a row, and the fourth is strawberry.

$\therefore$ P(selects 4 chocolates) $= \frac{7}{12} \times \frac{6}{11} \times \frac{5}{10} \times \frac{5}{9}$

≈ 0.0884

SOLUTIONS TO TOPIC 4 (STATISTICAL APPLICATIONS)

SHORT QUESTIONS

1 $p =$ P(first serve in) $= \frac{7}{9}$

The expected number of first serves in is $np = 180 \times \frac{7}{9}$

$= 140$

2

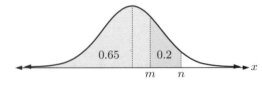

Using inverse normals:

P$(X \leqslant m) = 0.65$

$\therefore\ m \approx 45.7$

P$(X < n) = 0.65 + 0.2 = 0.85$

$\therefore\ n \approx 48.6$

$\therefore\ m - n = 48.6 - 45.7$

$= 2.9$

3 a

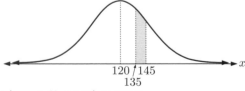

P$(135 \leqslant X \leqslant 145) \approx 0.116$

b

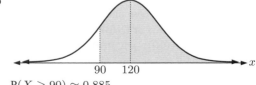

P$(X \geqslant 90) \approx 0.885$

c

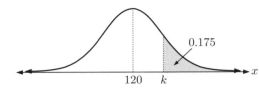

P$(X < k) = 0.825$

Using the inverse normal, $k \approx 143$

4 a P(two heads and one tail)

$=$ P(HHT or HTH or THH)

$=$ P(HHT) $+$ P(HTH) $+$ P(THH)

$= (\frac{1}{2})^3 + (\frac{1}{2})^3 + (\frac{1}{2})^3$

$= \frac{3}{8}$

b Expectation of exactly one tail

$= np$

$= 400 \times \frac{3}{8}$

$= 150$

We expect to see exactly one tail on 150 occasions.

5 a

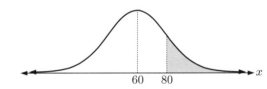

P$(X > 80) \approx 0.478$

b

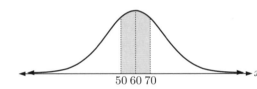

P$(50 < X < 70) \approx 0.595\,34$

We expect $90 \times 0.595\,34 \approx 54$ students to score between 50 and 70.

c

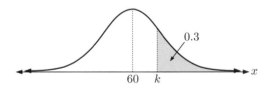

P$(X < k) = 0.7$

Using the inverse normal, $k \approx 66.3$.

We expect the minimum score to be about 66.3.

6 a $y \approx 1.09x - 0.781$ {using technology}

b $r \approx 0.883$

c There is a strong positive correlation between the times taken to complete the obstacle course of mice without adrenaline and the times of mice with adrenaline.

d Subject D is an outlier.

e $y \approx 1.094x - 0.7813$

Substituting in $y = 48$ gives

$48 \approx 1.094x - 0.7813$

$\therefore\ 48.7813 \approx 1.094x$

$\therefore\ x = 44.6$

So, we expect that an adrenaline injection would reduce the mouse's time to 44.6 s, a reduction of about 3.4 seconds.

7 a The 2×2 contingency table is:

	Y_1	Y_2	sum
X_1	32	14	46
X_2	25	19	44
sum	57	33	90

The expected frequency table is:

	Y_1	Y_2
X_1	$\dfrac{46 \times 57}{90} \approx 29$	$\dfrac{46 \times 33}{90} \approx 17$
X_2	$\dfrac{44 \times 57}{90} \approx 28$	$\dfrac{44 \times 33}{90} \approx 16$

b Using technology, $\chi^2_{calc} \approx 1.72$

8 a i

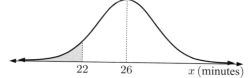

$P(X < 22) \approx 0.159$

$\therefore$ we expect $200 \times 0.159 \approx 32$ runners to have taken less than 22 minutes.

ii

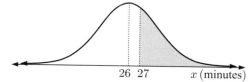

$P(X > 27) \approx 0.401$

$\therefore$ we expect $200 \times 0.401 \approx 80$ runners to have taken more than 27 minutes.

b

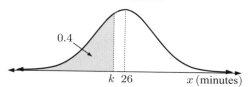

$P(X < k) = 0.4$

Using the inverse normal, $k \approx 25.0$

The fastest 40% of runners finished quicker than 25.0 minutes.

9 Expectation $= \displaystyle\sum_{i=1}^{n} x_i p_i$

$= 0 \times 0.19 + 1 \times 0.34 + 2 \times 0.38 + 3 \times 0.09$

$= 1.37$ points

10 a $r \approx 0.898$

b There is a strong correlation between the age of contestants and the time taken to complete the task.

c $y \approx 0.930x - 1.56$

d We expect that an increase of one year in age will add 0.93 minutes to the time to complete the task.

11 a H_0: *Preferred milk flavour* and *age* are independent.
H_1: *Preferred milk flavour* and *age* are dependent.

b Using technology, $\chi^2_{calc} \approx 10.6$

c $\chi^2_{calc} > \chi^2_{crit}$, so we reject the null hypothesis. We conclude at the 5% significance level that the preferred milk flavour is not independent of age.

12 a

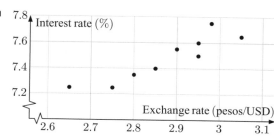

b $r \approx 0.915$

c There is a strong, positive, linear relationship between the exchange rate and interest rate.

13 a mean $\mu = 17$
standard deviation $\sigma = 3$

b

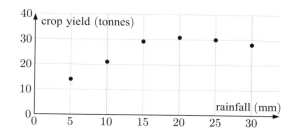

$P(X > 21) \approx 0.0912$

c The normal distribution is symmetric about the mean.

$\therefore \quad P(X > 22)$
$= P(X > \mu + 5)$
$= P(X < \mu - 5)$
$= P(X < 12)$
$\therefore \quad k = 12$

14 a H_0: Plant height is independent of light conditions.
H_1: Plant height is not independent of light conditions.

b Using technology, $\chi^2_{calc} \approx 1.59$

c Since $\chi^2_{calc} < \chi^2_{crit}$, we do not reject H_0.
The nursery's claim is justified according to this data. There is no significant difference in the height of the plants due to the conditions they are growing under.

15 a $r \approx 0.795$

b A moderate, positive relationship may exist between crop yield and rainfall.

c

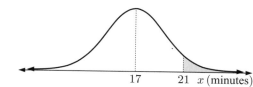

d The relationship between rainfall and crop yield does not appear to be linear and so r may not be appropriate for this data.

16 a $X \sim N(3.4, 0.3^2)$

i

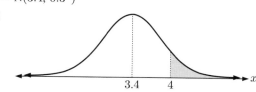

$P(X > 4) \approx 0.0228$

About 2.3% of babies weigh more than 4 kg.

ii

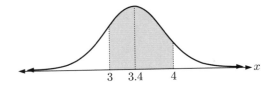

$P(3 < X < 4) \approx 0.886$
About 88.6% of babies weigh between 3 and 4 kg.

b Let k represent the maximum weight of *low birth weight* babies.

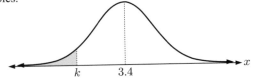

$P(X \leqslant k) = 0.1$
Using the inverse normal, $k \approx 3.02$.
10% of babies weigh less than 3.02 kg.

17 a $X \sim N(12, 3^2)$

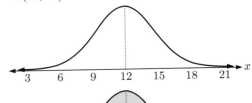

b

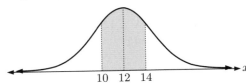

$P(10 \leqslant X \leqslant 14) \approx 0.495$

c i

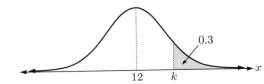

ii $P(X > k) = 0.3$
Using the inverse normal, $k \approx 13.6$

18 a i   P(globe is not faulty)
= 100% − P(globe is faulty)
= 100% − 0.5%
= 99.5%

ii   P(consecutive globes faulty)
= 0.005 × 0.005
= 0.000 025
= 0.0025%

b The company would expect $3000 \times 0.05 = 15$ globes to be faulty.

19 a mean $\mu = p$
standard deviation $\sigma = 1$

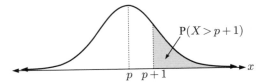

b The normal distribution is symmetric about the mean.
∴ $P(X > p + 1) = P(X < p - 1)$
∴ $k = p - 1$

20 H_0: The factors X and Y are independent.
H_1: The factors X and Y are dependent.
There are $(3 - 1) \times (3 - 1) = 4$ degrees of freedom.
Using technology, $\chi^2_{calc} \approx 6.88$.
The p-value ≈ 0.142.
Since $p > 0.05$, we do not reject H_0. We accept at the 5% level of significance, that the factors X and Y are independent.

21 a $y \approx -0.185x + 44.1$
b i $y \approx -0.185(100) + 44.1 \approx 26$
ii $y \approx -0.185(200) + 44.1 \approx 7$
c The first calculation is likely to be more reliable, as it is an interpolation. 200 grams is outside the range, so the second calculation is an extrapolation.

22 a i   P(both score at least one goal)
= P(Blake scores) × P(Miles scores)
= 0.15 × 0.25
= 0.0375

ii   P(at least one scores a goal)
= 1 − P(neither scores a goal)
= 1 − P(Blake misses) × P(Miles misses)
= 1 − 0.75 × 0.85
= 1 − 0.6375
= 0.3625

b From **a ii**, P(neither scores) = 0.6375
∴ in 15 games, we expect there to be $0.6375 \times 15 \approx 10$ games in which neither boy scored.

23 a The expected return
$= \frac{1}{12} \times \$40 + \frac{3}{12} \times \$20 + \frac{8}{12} \times \$5$
$\approx \$11.67$

b The expected return is less than the cost of playing the game.
∴ the game is not fair, and it is not advisable to play the game.

c Suppose x extra red tickets are added.
For the game to be fair, the expected return must be $15.
$\therefore \frac{1}{12 + x} \times \$40 + \frac{3 + x}{12 + x} \times \$20 + \frac{8}{12 + x} \times \$5 = \$15$
$\therefore \frac{40 + 20(3 + x) + 40}{12 + x} = 15$
$\therefore 140 + 20x = 15(12 + x)$
$\therefore 140 + 20x = 180 + 15x$
$\therefore 5x = 40$
$\therefore x = 8$

8 red tickets must be added.

24 a $y \approx 2.43x + 32.0$
b If $y = 70$, $70 \approx 2.43x + 32$
$\therefore 38 \approx 2.43x$
$\therefore x \approx 15.6$

So, Tony revised for 15.6 hours.

c The y-intercept (32%) is the estimate of the result for a student who did not do any revision.
The gradient of the line indicates that the result will increase by 2.43% for each additional hour studied.

25 a 3×3 contingency table:

satisfaction of employees

		high	medium	low	sum
quantity of ice cream	high	10	4	4	18
	low	14	20	10	44
	zero	64	50	54	168
	sum	88	74	68	230

Expected values:

satisfaction of employees

		high	medium	low
quantity of ice cream	high	$\frac{18 \times 88}{230} \approx 6.89$	$\frac{18 \times 74}{230} \approx 5.79$	$\frac{18 \times 68}{230} \approx 5.32$
	low	$\frac{44 \times 88}{230} \approx 16.8$	$\frac{44 \times 74}{230} \approx 14.2$	$\frac{44 \times 68}{230} \approx 13.0$
	zero	$\frac{168 \times 88}{230} \approx 64.3$	$\frac{168 \times 74}{230} \approx 54.1$	$\frac{168 \times 68}{230} \approx 49.7$

b H_0: Satisfaction of employees is independent of the quantity of supplied ice cream.

H_1: Satisfaction of employees is not independent of the quantity of supplied ice cream.

Using technology, $\chi^2_{calc} \approx 6.56$

Since $\chi^2_{calc} < \chi^2_{crit}$, we do not reject H_0.

We conclude at the 1% level of significance that the variables *quantity of ice cream* and *employee satisfaction* are independent.

26 $X \sim N(45, 4^2)$

a i

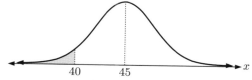

$P(X < 40) \approx 0.106$

$\therefore$ P(completes in < 40 seconds)
≈ 0.106

ii P(completes in < 40 seconds in two consecutive runs)
$\approx 0.106^2$
≈ 0.0112

b

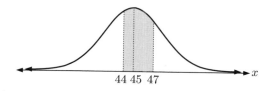

$P(44 < X < 47) \approx 0.2902$

$\therefore$ in 60 independent runs, we expect about $60 \times 0.2902 \approx 17$ runs to take between 44 and 47 seconds.

27 a $r \approx 0.993$

b There is a strong, positive linear correlation between h and t. As t increases, h increases also.

c i When $t = 14$, $h = 0.4879 \times 14 + 4.9145$
≈ 11.7

$\therefore$ the grass is about 11.7 mm high.

ii If $h = 20$, $20 = 0.4879t + 4.9145$
$\therefore$ $0.4879t = 15.0855$
$\therefore$ $t \approx 30.9$

It will take about 31 days for the grass to be 20 mm high.

28 a

	Painting	Sketching	Sculpting	sum
Male	30	35	15	80
Female	20	15	25	60
sum	50	50	40	140

The expected number of male sculptors is $\frac{80 \times 40}{140} \approx 23$.

b Using technology, $\chi^2_{calc} \approx 9.84$.

c Since $\chi^2_{calc} > \chi^2_{crit}$, we reject the null hypothesis that *gender* and *choice of art speciality* are independent. We conclude at the 1% level that *gender* and *choice of art speciality* are dependent.

LONG QUESTIONS

1 $X \sim N(254, 2.3^2)$

a $P(X < 254) = P(X < \mu) = 0.5$

b

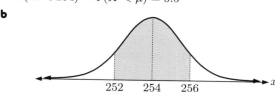

$P(252 < X < 256) \approx 0.615$
$\approx 61.5\%$

c

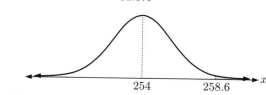

$P(X > 254 + 2\sigma) = P(X > 258.6)$
≈ 0.0228

The expected number of drinks with volume at least two standard deviations above the mean is $\approx 80 \times 0.0228$
≈ 2

d i

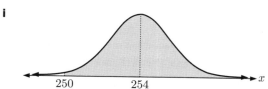

$P(X \geqslant 250) \approx 0.959$

$\therefore$ the guarantee is valid.

ii The new distribution is $X \sim N(254, 2.5^2)$.

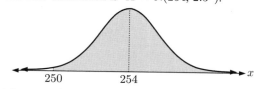

$P(X \geqslant 250) \approx 0.945$

$\therefore$ the guarantee is now invalid.

2 a, b

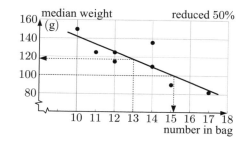

c As the number of tomatoes in a bag increases, the median weight of tomatoes in the bag decreases.

d **i** median weight is approximately 118 g

    ii 15 tomatoes per bag

e **i** $y \approx -8.21x + 224$

    ii $y \approx -8.21(20) + 224 \approx 59.8$ g

    iii This is an extrapolated value, and therefore not reliable.

    iv $r \approx -0.832$

3 **a** H_0: orange yield is independent of use of new fertiliser.

    H_1: orange yield is not independent of use of new fertiliser.

b Expected yield of Variety A oranges with new fertiliser $= \frac{188 \times 105}{340} \approx 58.1$

c Using technology, $\chi^2_{calc} \approx 4.72$

d df $= (2-1)(3-1) = 2$ **e** p-value ≈ 0.0944

f Since the p-value $\approx 0.0944 > 0.05$, we do not reject the null hypothesis. We thus conclude that orange yield is independent of the fertiliser type, at a 5% significance level.

4 **a**

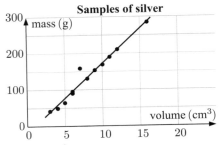

b **i** $\overline{x} = \dfrac{3 + 6 + 4 + \dots + 10 + 11}{12} \approx 8.08$ cm³

      $\overline{y} = \dfrac{40 + 95 + 50 + \dots + 170 + 190}{12} \approx 137$ g

    ii on graph

c $r \approx 0.980$

d A very strong, positive relationship appears to exist between volume and mass of the samples of silver.

e Yes, sample D is well away from the general trend.

f **i** $y \approx 19.5x - 24.7$

    ii $y \approx 19.5(7) - 24.7 \approx 112$ g

    iii Percentage error between the given and expected mass is $\dfrac{160 - 112}{112} \times 100\% \approx 42.9\%$.

5 **a**

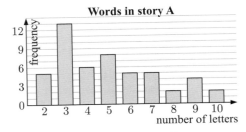

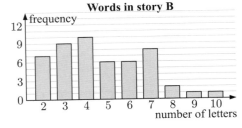

b

	A	B
min	2	2
Q_1	3	3
med	5	4
Q_3	7	6
max	10	10

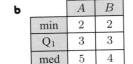

c A: mean $= 5$, standard deviation ≈ 2.28

   B: mean $= 4.76$, standard deviation ≈ 2.02

d

	< 5	$\geqslant 5$	sum
Story A	24	26	50
Story B	26	24	50
sum	50	50	100

e H_0: There is no difference in the stories.

   H_1: There is some difference in the stories.

   Using technology, $\chi^2_{calc} \approx 0.16$

   Since $\chi^2_{calc} < \chi^2_{crit}$, we do not reject H_0.

f There is no significant difference in word length between the stories.

g The graphical evidence suggests that the stories may have been written by different authors.

   The mean of B is lower and the standard deviation is smaller, suggesting that the stories are by different authors. However, the chi-squared test suggests that the difference is not significant.

   Overall, the statistical evidence is not strong enough to support the claim that the two stories were written by different authors.

6 **a** H_0: Choice of breakfast cereal is independent of gender.

b Contingency table:

	Muesli	Rolled Oats	Corn Flakes	Weetbix	sum
Female	4	14	48	24	90
Male	8	30	26	26	90
sum	12	44	74	50	180

Expected values:

	Muesli	Rolled Oats	Corn Flakes	Weetbix
Female	$\frac{90 \times 12}{180}$ $= 6$	$\frac{90 \times 44}{180}$ $= 22$	$\frac{90 \times 74}{180}$ $= 37$	$\frac{90 \times 50}{180}$ $= 25$
Male	$\frac{90 \times 12}{180}$ $= 6$	$\frac{90 \times 44}{180}$ $= 22$	$\frac{90 \times 74}{180}$ $= 37$	$\frac{90 \times 50}{180}$ $= 25$

c Using technology, $\chi^2_{calc} \approx 13.8$

d Since $\chi^2_{calc} > \chi^2_{crit}$, we reject H_0.

e Jack can conclude that, at a 5% significance level, *choice of breakfast cereal* and *gender* are dependent.

7 **a**

Age	Income ($000s)
17	13
19	15
21	16
23	18
23	20
25	18
26	21
32	24
37	27
40	28

b $r \approx 0.979$

c There is a strong, positive correlation between *age* and *annual income*.

d $y \approx 0.644x + 3.05$

e **i** When $x = 30$, $y \approx 0.644(30) + 3.05$
≈ 22.4
$\therefore$ the estimated income for a 30 year old is \$22 400.

ii When $x = 60$, $y \approx 0.644(60) + 3.05$
≈ 41.7
$\therefore$ the estimated income for a 60 year old is \$41 700.

f **e i** is an interpolation, so it is likely to be reliable.

e ii is an extrapolation, so it may be unreliable.

8 **a**

Die 2

		1	2	3	4
Die 1	1	2	3	4	5
	2	3	4	5	⑥
	3	4	5	⑥	⑦
	4	5	⑥	⑦	⑧

b **i** $P(S = 5) = \frac{4}{16} = \frac{1}{4}$

ii $P(S > 5) = \frac{6}{16} = \frac{3}{8}$

iii $P(S = 5 \mid S > 3) = \dfrac{P(S = 5)}{P(S > 3)} = \frac{4}{13}$

c **i** The expected number of points
$= 32 \times \frac{1}{16} + 16 \times \frac{9}{16} + (-8) \times \frac{6}{16}$
$= 2 + 9 + (-3)$
$= 8$

ii Let k be the number of points lost for $S > 5$.
For the expectation to be zero,
$32 \times \frac{1}{16} + 16 \times \frac{9}{16} - k \times \frac{6}{16} = 0$
$\therefore \frac{6k}{16} = 11$
$\therefore k \approx 29.3$
So, ≈ 29.3 points must be lost for $S > 5$.

9 **a** **i** H_0: age and reaction time are independent.

ii H_1: age and reaction time are not independent.

b **i** 64 were teenagers

ii 66 had a slow reaction time

c The expected number of teenagers with a slow reaction
time $= \dfrac{66 \times 64}{200}$
$= 21.12$

d df $= (3 - 1)(4 - 1) = 6$

e $\chi^2_{calc} \approx 13.2$

f Since $\chi^2_{calc} > \chi^2_{crit}$, we reject the null hypothesis and conclude that age and reaction time are dependent.

g At a 1% significance level, $\chi^2_{calc} < \chi^2_{crit}$.
In this case the researcher would not reject the null hypothesis, and so conclude that age and reaction time were independent.

SHORT QUESTIONS

1 **a**

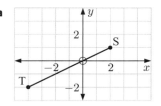

b ST $= \sqrt{(-4 - 2)^2 + (-2 - 1)^2}$ {distance formula}
$= \sqrt{36 + 9}$
$= \sqrt{45} \approx 6.71$ units

c SR $= 5$ units
$\therefore \sqrt{(b - 2)^2 + (4 - 1)^2} = 5$ {distance formula}
$\therefore (b - 2)^2 + 9 = 25$
$\therefore (b - 2)^2 = 16$
$\therefore b - 2 = \pm 4$
$\therefore b = 6$ {as $b > 0$}

2

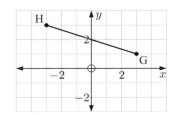

a Gradient of GH $= \dfrac{1 - 3}{3 - -3} = \dfrac{-2}{6} = -\frac{1}{3}$

b Midpoint of GH is $\left(\dfrac{3 + -3}{2}, \dfrac{1 + 3}{2} \right)$ which is $(0, 2)$.

c The perpendicular bisector has gradient 3, so it has the form $y = 3x + b$.
It passes through $(0, 2)$, so $2 = 3(0) + b$
$\therefore b = 2$
$\therefore$ the equation is $y = 3x + 2$.

3 **a** L_1: $4x - 3y = 7$
$\therefore -3y = -4x + 7$
$\therefore y = \dfrac{-4x}{-3} + \dfrac{7}{-3}$
$\therefore y = \frac{4}{3}x - \frac{7}{3}$
$\therefore$ gradient of $L_1 = \frac{4}{3}$

b L_2: $2x + ky = 5$
$\therefore ky = -2x + 5$
$\therefore y = -\dfrac{2}{k}x + \dfrac{5}{k}$
$\therefore$ gradient of $L_2 = -\dfrac{2}{k}$

i If L_1 and L_2 are parallel, $\dfrac{-2}{k} = \dfrac{4}{3}$
$\therefore 4k = -6$
$\therefore k = \dfrac{-6}{4} = -\frac{3}{2}$

ii If L_1 and L_2 are perpendicular, $-\dfrac{2}{k} \times \dfrac{4}{3} = -1$
$\therefore -8 = -3k$
$\therefore \frac{8}{3} = k$

Mathematical Studies SL – Exam Preparation & Practice Guide (3rd edition)

4 a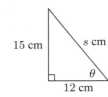

$\tan \theta = \dfrac{\text{OPP}}{\text{ADJ}} = \dfrac{15}{12} = \dfrac{5}{4}$

$\therefore \quad \theta \approx 51.3°$

b $s^2 = 12^2 + 15^2$ {Pythagoras}

$\therefore \quad s^2 = 369$

$\therefore \quad s \approx 19.2$ {since $s > 0$}

c surface area $= \pi rs + \pi r^2$

$= \pi(12)\sqrt{369} + \pi(12)^2$

$\approx 1180 \text{ cm}^2$

5 a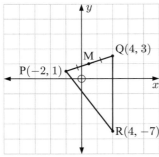

b M has coordinates $\left(\dfrac{-2+4}{2}, \dfrac{1+3}{2}\right)$ or $(1, 2)$.

c The circle has radius MR $= \sqrt{(4-1)^2 + (-7-2)^2}$

$= \sqrt{90}$ units

$\therefore$ its area $= \pi r^2$

$= 90\pi$

$\approx 283 \text{ units}^2$

6 a The gradient is $\dfrac{y_2 - y_1}{x_2 - x_1} = \dfrac{10-4}{-1-(-3)} = \dfrac{6}{2} = 3$

b The equation of the line is $y = 3x + c$.

The line passes through $(-3, 4)$, so $4 = 3(-3) + c$

$\therefore \quad c = 13$

$\therefore$ the equation is $y = 3x + 13$.

c When $x = 0$, $y = 13$ $\therefore$ the y-intercept is 13.

When $y = 0$, $3x + 13 = 0$

$\therefore \quad x = -\dfrac{13}{3}$

$\therefore$ the x-intercept is $-\dfrac{13}{3}$.

7 a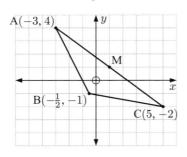

The longest side is clearly from A$(-3, 4)$ to C$(5, -2)$.

Its midpoint is $\left(\dfrac{-3+5}{2}, \dfrac{4-2}{2}\right)$ or $(1, 1)$.

b Distance BM

$= \sqrt{(1-(-\frac{1}{2}))^2 + (1-(-1))^2}$ {distance formula}

$= \sqrt{6.25}$

$= 2.5$ units

c Since $\triangle$ABC is isosceles, BM is perpendicular to AM.

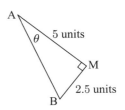

$\tan \theta = \dfrac{\text{OPP}}{\text{ADJ}}$

$= \dfrac{2.5}{5} = \dfrac{1}{2}$

$\therefore \quad \theta \approx 26.6°$

8 a $3x + 2y = 5$

$\therefore \quad 2y = 5 - 3x$

$\therefore \quad y = -\dfrac{3}{2}x + \dfrac{5}{2}$

$\therefore \quad L_1$ has gradient $-\dfrac{3}{2}$. $\therefore \quad L_2$ has gradient $\dfrac{2}{3}$.

b L_2 has equation $y = \dfrac{2}{3}x + c$

It passes through $(2, -1)$, so $-1 = \dfrac{4}{3} + c$

$\therefore \quad c = -\dfrac{7}{3}$

So, $y = \dfrac{2}{3}x - \dfrac{7}{3}$

c When $y = 0$, $\dfrac{2}{3}x - \dfrac{7}{3} = 0$

$\therefore \quad \dfrac{2}{3}x = \dfrac{7}{3}$

$\therefore \quad 2x = 7$

$\therefore \quad x = \dfrac{7}{2}$

$\therefore$ the x-intercept is $\dfrac{7}{2}$.

9 a Let the distance be d m.

$\sin 35° = \dfrac{\text{OPP}}{\text{HYP}} = \dfrac{d}{8}$

$\therefore \quad d = 8 \sin 35°$

$\therefore \quad d \approx 4.5886$

The foot is about 4.59 m from the base of the wall.

b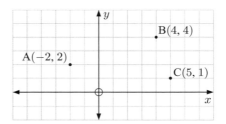

$\cos \theta = \dfrac{\text{ADJ}}{\text{HYP}}$

$\approx \dfrac{6.5886}{8}$

$\therefore \quad \theta \approx 34.6°$

10 a

b $m_{AB} = \dfrac{4-2}{4-(-2)} = \dfrac{2}{6} = \dfrac{1}{3}$

$m_{BC} = \dfrac{1-4}{5-4} = \dfrac{-3}{1} = -3$

$m_{AB} \times m_{BC} = -1$, so AB $\perp$ BC

So, the triangle is right angled at B.

c AC $= \sqrt{(5-(-2))^2 + (1-2)^2}$ {distance formula}

$= \sqrt{50} \approx 7.07$ units

11 a $L_1: \quad 7x + 3y = 12$

$\therefore \quad 3y = -7x + 12$

$\therefore \quad y = -\dfrac{7}{3}x + 4$

$\therefore$ the gradient of $L_1 = -\dfrac{7}{3}$

$\therefore \quad L_2$ has gradient $-\dfrac{7}{3}$ {parallel lines}

b A perpendicular line has gradient $= \frac{3}{7}$
$$\therefore \quad y = \frac{3}{7}x + c$$
But $(7, 1)$ lies on the line.
$$\therefore \quad 1 = \frac{3}{7}(7) + c$$
$$\therefore \quad 1 = 3 + c$$
$$\therefore \quad c = -2$$
$\therefore$ the equation is $y = \frac{3}{7}x - 2$.

12 a $2x - 4y = 7$ meets the x-axis when $x = k$ and $y = 0$
$$\therefore \quad 2k = 7$$
$$\therefore \quad k = \frac{7}{2}$$

b The line passing through $(3, 2)$ and $(-6, 0)$ has
$$\text{gradient} = \frac{0 - 2}{-6 - 3} = \frac{-2}{-9} = \frac{2}{9}$$
Its equation is $y = \frac{2}{9}x + c$.
But $(-6, 0)$ lies on the line.
$$\therefore \quad 0 = \frac{2}{9}(-6) + c$$
$$\therefore \quad 0 = -\frac{12}{9} + c$$
and so $c = \frac{12}{9}$
So, the line is $y = \frac{2}{9}x + \frac{12}{9}$
$$\therefore \quad 9y = 2x + 12$$
$$\therefore \quad 2x - 9y + 12 = 0$$
$$\therefore \quad a = 2, \ b = -9, \ d = 12$$

13 a

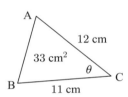

b Let $\widehat{ACB} = \theta$
$$\therefore \quad \frac{1}{2} \times 11 \times 12 \times \sin\theta = 33$$
$$\therefore \quad \sin\theta = \frac{1}{2}$$
$$\therefore \quad \theta = 30° \quad \{\theta \text{ is acute}\}$$
and so $\widehat{ACB} = 30°$

c Using the cosine rule,
$$AB^2 = AC^2 + BC^2 - 2 \times AC \times BC \times \cos\theta$$
$$\therefore \quad AB = \sqrt{12^2 + 11^2 - 2(12)(11)\cos 30°}$$
$$\approx 6.03 \text{ cm}$$

14 a $P(3, k)$ lies on L $\therefore$ $k = 3 - 2 \times 3$
$$\therefore \quad k = -3$$

b L has gradient -2.

c The perpendicular to L has gradient $\frac{1}{2}$, so its equation is $y = \frac{1}{2}x + c$.
The perpendicular passes through $P(3, -3)$
so $\frac{1}{2} \times 3 + c = -3$
$$\therefore \quad c = -3 - \frac{3}{2} = -\frac{9}{2}$$
The line is $y = \frac{1}{2}x - \frac{9}{2}$.

15 a

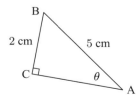

b Let $\widehat{BAC} = \theta$
$$\therefore \quad \sin\theta = \frac{\text{OPP}}{\text{HYP}} = \frac{2}{5}$$
$$\therefore \quad \theta \approx 23.58° \text{ and so } \widehat{BAC} \approx 23.58° \ \{\text{to 4 s.f.}\}$$

c $AC^2 + 2^2 = 5^2$ $\{$Pythagoras$\}$
$$\therefore \quad AC^2 = 25 - 4 = 21$$
$$\therefore \quad AC = \sqrt{21}$$
$$\approx 4.58 \text{ cm} \quad \{\text{to 3 s.f.}\}$$

16

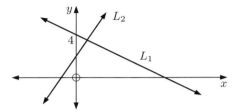

a L_1 has gradient $= -\frac{1}{2}$ and y-intercept $= 4$
$$\therefore \quad L_1 \text{ is } y = -\frac{1}{2}x + 4$$
which is $2y = -x + 8$
or $x + 2y - 8 = 0$

b Gradient of $L_2 = \frac{8 - -1}{4 - -2} = \frac{9}{6} = \frac{3}{2}$
$$\therefore \quad y = \frac{3}{2}x + c$$
But L_2 passes through $(-2, -1)$
$$\therefore \quad -1 = \frac{3}{2}(-2) + c$$
$$\therefore \quad -1 = -3 + c \text{ and so } c = 2$$
So, L_2 is $y = \frac{3}{2}x + 2$.
L_1 and L_2 intersect when
$$-\frac{1}{2}x + 4 = \frac{3}{2}x + 2$$
$$\therefore \quad 2 = 2x \text{ and so } x = 1$$
Substituting $x = 1$ into $y = \frac{3}{2}x + 2$
$$\text{gives } y = \frac{3}{2}(1) + 2 = 3\frac{1}{2}$$
$\therefore$ the point of intersection is $(1, 3\frac{1}{2})$.

17 a $kx + 3y = 7$ cuts the x-axis at $(4, 0)$
$$\therefore \quad 4k = 7$$
$$\therefore \quad k = \frac{7}{4}$$

b $(1, -1)$ lies on $ax - y = 4h$
$$\therefore \quad a + 1 = 4h \quad .... \ (1)$$
$(1, -1)$ also lies on $ax + 2y = h$
$$\therefore \quad a - 2 = h \quad .... \ (2)$$
Subtracting (2) from (1), $3 = 3h$
$$\therefore \quad h = 1$$
Substituting $h = 1$ into (1), $a + 1 = 4$
$$\therefore \quad a = 3$$

18 a Using technology, $y = \dfrac{24}{x}$ and $y = 3x - 1$ intersect at $(3, 8)$.

b The gradient of $y = 3x - 1$ is 3.
$\therefore$ the gradient of the perpendicular is $-\frac{1}{3}$.
$$\therefore \quad y = -\frac{1}{3}x + c$$
The perpendicular passes through $(3, 8)$,
so $8 = -\frac{1}{3}(3) + c$
$$\therefore \quad 8 = -1 + c$$
$$\therefore \quad 9 = c$$
$$\therefore \quad y = -\frac{1}{3}x + 9 \text{ is the perpendicular line.}$$

Mathematical Studies SL – Exam Preparation & Practice Guide (3ʳᵈ edition)

19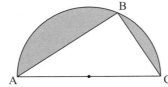

AB = 8 cm
BC = 6 cm
AC = 10 cm

a $AB^2 + BC^2 = 8^2 + 6^2 = 64 + 36$
$$= 100$$
$$AC^2 = 10^2 = 100$$
$$\therefore \quad AB^2 + BC^2 = AC^2$$
$\therefore \quad \triangle ABC$ is right angled at B.

b Shaded area = area semi-circle − area triangle
$$= \tfrac{1}{2} \times \pi \times 5^2 - \tfrac{1}{2} \times 8 \times 6$$
$$\approx 15.3 \text{ cm}^2$$

20 a When $x = 0$, $2y - 6 = 0$
$$\therefore \quad y = 3 \quad \therefore \text{ A is } (0, 3)$$
When $y = 0$, $3x - 6 = 0$
$$\therefore \quad 3x = 6$$
$$\therefore \quad x = 2 \quad \therefore \text{ B is } (2, 0)$$

b $3x + 2y - 6 = 0$
$$\therefore \quad 2y = -3x + 6$$
$$\therefore \quad y = -\tfrac{3}{2}x + 3$$
$\therefore \quad$ the gradient is $-\tfrac{3}{2}$.

c The gradient of M is $\tfrac{2}{3}$ so its equation is $y = \tfrac{2}{3}x + c$.
But M passes through A, so $c = 3$.
$\therefore \quad$ the equation is $y = \tfrac{2}{3}x + 3$.

21 a
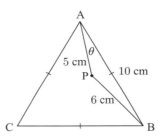

b **i** Let $\widehat{BAP} = \theta$.
Using the cosine rule,
$$6^2 = 5^2 + 10^2 - 2 \times 5 \times 10 \times \cos\theta$$
$$\therefore \quad 100\cos\theta = 89$$
$$\therefore \quad \cos\theta = \tfrac{89}{100}$$
$$\therefore \quad \theta \approx 27.1° \quad \text{and so} \quad \widehat{BAP} \approx 27.1°$$

ii $\widehat{CAP} = 60° - \widehat{BAP}$
$$\approx 32.9°$$

c Using the cosine rule,
$$CP^2 \approx 10^2 + 5^2 - 2 \times 5 \times 10 \times \cos 32.9°$$
$$\approx 41.01$$
$$\therefore \quad CP \approx 6.40 \text{ cm}$$

22 a
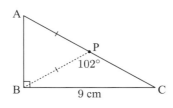

b $2\,\widehat{BAP} = 102°$ {exterior angle of $\triangle$}
$$\therefore \quad \widehat{BAP} = 51°$$
Now $\sin 51° = \dfrac{\text{OPP}}{\text{HYP}} = \dfrac{9}{AC}$

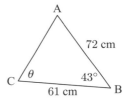

$$\therefore \quad AC = \dfrac{9}{\sin 51°}$$
$$\approx 11.6 \text{ cm}$$

23 a

2:35 pm   2:38 pm   2:42 pm

$25°$ θ

h m

Suppose the plane has altitude h m.
The distance travelled between 2:35 pm and 2:38 pm
is $3 \times 60 \times 110 = 19\,800$ m.
Now $\tan 25° = \dfrac{\text{OPP}}{\text{ADJ}} = \dfrac{h}{19\,800}$
$$\therefore \quad h = 19\,800 \tan 25°$$
$$\therefore \quad h = 9233 \text{ m}$$
$\therefore \quad$ the plane is about 9230 m above the ground.

b Suppose the angle of elevation of the plane at 2:42 pm
is θ.
The distance travelled between 2:35 pm and 2:42 pm
is $7 \times 60 \times 110 = 46\,200$ m
Now $\tan\theta = \dfrac{\text{OPP}}{\text{ADJ}} = \dfrac{h}{46\,200} \approx \dfrac{9233}{46\,200}$
$$\therefore \quad \theta \approx 11.3°$$
$\therefore \quad$ the angle of elevation of the plane is about $11.3°$.

24 a

A

72 cm

θ $43°$

C   61 cm   B

Using the cosine rule,
$$(AC)^2 = 61^2 + 72^2 - 2 \times 61 \times 72 \times \cos 43°$$
$$\approx 2480.8$$
$$\therefore \quad AC \approx 49.808$$
$$\approx 49.8 \text{ cm} \quad \{\text{as } AC > 0\}$$

b Let $\widehat{ACB} = \theta$
Using the sine rule, $\dfrac{\sin\theta}{72} = \dfrac{\sin 43°}{AC}$
$$\therefore \quad \sin\theta \approx \dfrac{72 \sin 43°}{49.808}$$
$$\approx 0.9859$$
$$\therefore \quad \theta \approx 80.4°$$
and so $\widehat{ACB} \approx 80.4°$

25 a

P

h m

$36°$

Q   40 m   B

Let the height of the pole
be h m.
$\tan 36° = \dfrac{\text{OPP}}{\text{ADJ}} = \dfrac{h}{40}$
$$\therefore \quad h = 40 \tan 36°$$
$$\therefore \quad h \approx 29.062$$

So, the pole is about 29.1 m high.

b

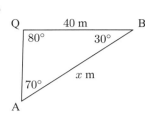

$A\widehat{Q}B = 180° - 30° - 70°$
$= 80°$

Let $AB = x$ m.
Using the sine rule,

$$\frac{x}{\sin 80°} = \frac{40}{\sin 70°}$$

$$\therefore \quad x = \frac{40\sin 80°}{\sin 70°}$$

$$\approx 41.92$$

$\therefore$ AB is about 41.9 m.

26 a

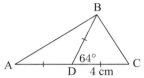

$\triangle$BCD is isosceles, so $2 \times D\widehat{B}C + 64° = 180°$

$$\therefore \quad 2 \times D\widehat{B}C = 116°$$

$$\therefore \quad D\widehat{B}C = 58°$$

$\triangle$ABD is isosceles, so $2A\widehat{B}D = 64°$

$$\therefore \quad A\widehat{B}D = 32°$$

$$\therefore \quad A\widehat{B}C = A\widehat{B}D + D\widehat{B}C$$

$$= 90°$$

b

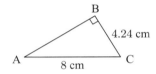

$(AB)^2 = 8^2 - 4.24^2$
{Pythagoras}
$= 46.0224$

$\therefore$ AB ≈ 6.78 cm

c Area $= \frac{1}{2} \times AB \times BC$

$$\approx \frac{1}{2} \times 6.783\,98 \times 4.24$$

$$\approx 14.4 \text{ cm}^2$$

27

a $A\widehat{D}E = 110° - 90° = 20°$
Let AE be x cm.

$$\therefore \quad \tan 20° = \frac{x}{40}$$

$$\therefore \quad x = 40\tan 20°$$

$$\approx 14.56$$

$$\therefore \quad AB \approx 14.56 + 7$$

$$\approx 21.56 \text{ cm} \quad (2 \text{ dec. pl.})$$

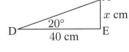

b $\cos 20° = \dfrac{40}{AD}$

$$\therefore \quad AD = \frac{40}{\cos 20°}$$

$$\approx 42.6 \text{ cm}$$

c ABCD is a trapezium with DC parallel to AB.

$$\text{area} = \frac{h}{2}(a + b)$$

$$\therefore \quad \text{area} \approx \frac{40}{2}(7 + 21.56)$$

$$\approx 571 \text{ cm}^2$$

28

a Area $\triangle$DEF $= 120$ cm^2. Let DF be x cm.
Using area of a triangle $= \frac{1}{2}bc\sin A$
we have $120 = \frac{1}{2} \times 12 \times x\sin 30°$

$$\therefore \quad 120 = 6 \times x \times \frac{1}{2}$$

$$\therefore \quad 120 = 3x$$

$$\therefore \quad 40 = x$$

$$\therefore \quad DF = 40 \text{ cm}$$

b In $\triangle$EDG, $\sin 30° = \dfrac{EG}{12}$

$$\therefore \quad EG = 12\sin 30° = 6$$

The length of the perpendicular is 6 cm.

29 a

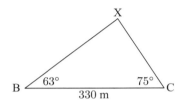

b $B\widehat{X}C = 180° - (63° + 75°) = 42°$

Using the sine rule, $\dfrac{BX}{\sin 75°} = \dfrac{330}{\sin 42°}$

$$\therefore \quad BX = \frac{330\sin 75°}{\sin 42°} \approx 476$$

The distance to the monument from B is 476 m.

30

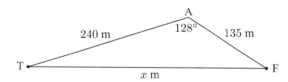

a Let the distance from T to F be x m.
Using the cosine rule,

$$x^2 = 240^2 + 135^2 - 2 \times 240 \times 135\cos 128°$$

$$\therefore \quad x \approx 340.18$$

So, the distance is about 340 m.

b Using the sine rule, $\dfrac{\sin A\widehat{T}F}{135} = \dfrac{\sin 128°}{340.18}$

$$\therefore \quad A\widehat{T}F \approx \sin^{-1}\left(\frac{135\sin 128°}{340.18}\right) \approx 18.2°$$

31 a

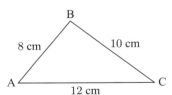

b The smallest angle is opposite the shortest side.
This is $B\widehat{C}A$, which we call θ.

Using the cosine rule, $\cos\theta = \dfrac{10^2 + 12^2 - 8^2}{2 \times 10 \times 12} = 0.75$

$$\therefore \quad \theta = \cos^{-1}(0.75) \approx 41.4°$$

c Area $\triangle$ABC $\approx \frac{1}{2} \times 10 \times 12\sin 41.4° \approx 39.7$ cm^2

32

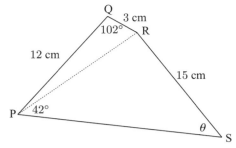

a Using the cosine rule,
$$PR^2 = 12^2 + 3^2 - 2 \times 12 \times 3 \cos 102° \approx 167.970$$
$$\therefore \quad PR \approx 12.960 \text{ cm, which is } \approx 13.0 \text{ cm}$$

b Using the sine rule, $\dfrac{\sin \theta}{PR} = \dfrac{\sin 42°}{15}$
$$\therefore \quad \sin \theta \approx \frac{12.960 \sin 42°}{15} \approx 0.578$$
$$\therefore \quad \theta \approx 35.3°$$

33

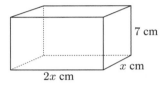

a Let the shorter side be x cm, so the longer side is $2x$ cm
$$V = 2x \times x \times 7 \text{ cm}$$
$$\text{But} \quad V = 350 \text{ cm}^3$$
$$\therefore \quad 14x^2 = 350$$
$$\therefore \quad x^2 = \frac{350}{14} = 25$$
$$\therefore \quad x = \sqrt{25} = 5 \quad \{\text{as } x > 0\}$$
$$\therefore \quad \text{dimensions are } 5 \text{ cm} \times 10 \text{ cm} \times 7 \text{ cm.}$$

b

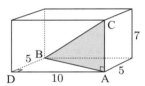

We need the length of the diagonal of the box.

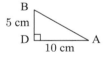

In $\triangle ABD$, $AB^2 = 10^2 + 5^2$ {Pythagoras}
$$\therefore \quad AB^2 = 125$$

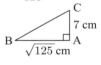

In $\triangle ABC$, $BC^2 = AB^2 + 7^2$ {Pythagoras}
$$\therefore \quad BC^2 = 125 + 49 = 174$$
$$\therefore \quad BC = \sqrt{174} \quad \{\text{as } BC > 0\}$$
$$\approx 13.2$$
$$\therefore \quad \text{the longest pencil is 13.2 cm long.}$$

34 a

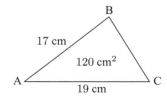

b Let $\widehat{BAC} = \theta$.

Using area $= \frac{1}{2}bc \sin A$,
$$120 = \frac{1}{2} \times 19 \times 17 \sin \theta$$
$$\therefore \quad \sin \theta = \frac{120}{161.5}$$
$$\therefore \quad \theta = \sin^{-1}\left(\frac{120}{161.5}\right) \approx 48.0°$$
So, $\widehat{BAC} \approx 48.0°$

c Using the cosine rule,
$$BC^2 \approx 17^2 + 19^2 - 2 \times 17 \times 19 \cos 48.0°$$
$$\therefore \quad BC \approx 14.8$$
So, the remaining side is 14.8 cm long.

d Volume = area cross-section × length
$$= 120 \times 13.5$$
$$= 1620 \text{ cm}^3$$

35 a

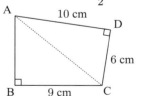

b Length of base diagonal
$$= \sqrt{50^2 + 50^2} \approx 70.7 \text{ m}$$

Length of sloping edge
$$\approx \sqrt{7^2 + \left(\frac{70.7}{2}\right)^2}$$
$$\approx 36.0 \text{ m}$$

c $\sin \theta \approx \dfrac{7}{36.0}$
$$\therefore \quad \theta \approx \sin^{-1}\left(\frac{7}{36.0}\right) \approx 11.2°$$
The angle between the sloping edge and the base diagonal is about 11.2°.

d

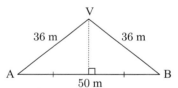

The height is $\sqrt{(36.0)^2 - 25^2} \approx 25.96$ m

The area of each face is $\dfrac{50 \times 25.96}{2} \approx 649 \text{ m}^2$

36

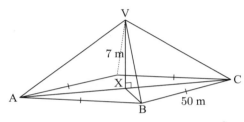

a In $\triangle ADC$, $AD^2 + DC^2 = AC^2$ {Pythagoras}
$$\therefore \quad 10^2 + 6^2 = AC^2$$
$$\therefore \quad AC^2 = 136$$

In $\triangle ABC$, $AB^2 + BC^2 = AC^2$ {Pythagoras}
$$\therefore \quad AB^2 + 9^2 = 136$$
$$\therefore \quad AB^2 = 136 - 81 = 55$$
$$\therefore \quad AB = \sqrt{55} \approx 7.42 \text{ cm}$$

b Area of quadrilateral ABCD
= area of $\triangle ABC$ + area of $\triangle ADC$

$$= \frac{\sqrt{55} \times 9}{2} + \frac{10 \times 6}{2}$$

≈ 63.37

$\approx 63.4 \text{ cm}^2$

c $\qquad$ Volume = area of cross-section $\times$ length

$\therefore$ length of prism = $\frac{484}{63.37} \approx 7.64$ cm

37 **a** Using the cosine rule,

$x^2 = 1200^2 + 2100^2 - 2 \times 1200 \times 2100 \times \cos 48°$

$\therefore$ $x \approx 1574$ {as $x > 0$}

The distance is 1574 m. {to the nearest metre}

b Using the sine rule, $\dfrac{\sin O\widehat{A}B}{1200} = \dfrac{\sin 48°}{x}$

$\therefore$ $\sin O\widehat{A}B \approx \dfrac{1200 \sin 48°}{1574}$

≈ 0.567

$\therefore$ $O\widehat{A}B \approx 34.5°$

c $\qquad$ Total distance $\approx 1200 + 2100 + 1574$

≈ 4874 m

Speed = 8 km h^{-1} = $\dfrac{8000}{60}$ m min^{-1}

$= \dfrac{400}{3}$ m min^{-1}

$\therefore$ time = $\dfrac{\text{distance}}{\text{speed}} \approx 4874 \times \dfrac{3}{400}$

≈ 36.6 minutes

So, it will take Jerry about 37 minutes.

38

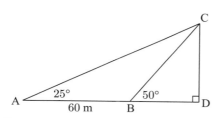

a $\qquad A\widehat{B}C = 180° - 50°$ {angles in a straight angle}

$= 130°$

$\therefore$ $A\widehat{C}B = 180° - (25° + 130°)$

$= 25°$

b $\triangle ACB$ is isosceles {base angles equal}

$\therefore$ BC = AB = 60 m

Let the tower have height x m.

$\therefore$ $\sin 50° = \dfrac{x}{60}$

$\therefore$ $x = 60 \sin 50°$

$\therefore$ $x \approx 46.0$

The tower is 46.0 m high.

39

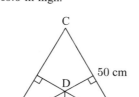

---

a $A\widehat{B}C = 60°$ {equilateral $\triangle$}

b Each altitude bisects the angle and the base.
This gives triangle

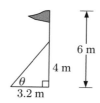

$\tan 30° = \dfrac{x}{25}$

$\therefore$ $x = 25 \tan 30°$

$\therefore$ $x \approx 14.4$

The shortest distance is about 14.4 cm.

40 **a**

b The wire is secured to the pole $\frac{2}{3}$ of the way up,
which is $\frac{2}{3} \times 6 = 4$ m.

Let θ represent the angle of elevation of each wire.

$\tan \theta = \dfrac{4}{3.2}$

$\therefore$ $\theta = \tan^{-1}\left(\dfrac{4}{3.2}\right) \approx 51.3°$

So, the angle of elevation of each wire is about 51.3°.

c Let each wire be x m long.

$4^2 + 3.2^2 = x^2$ {Pythagoras' theorem}

$\therefore$ $16 + 10.24 = x^2$

$\therefore$ $x^2 = 26.24$

$\therefore$ $x = \sqrt{26.24}$ {$x > 0$}

≈ 5.122

$\therefore$ the total length of the three wires $\approx 3 \times 5.122$

≈ 15.4 m

LONG QUESTIONS

1 **a**

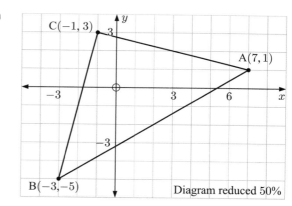

Diagram reduced 50%

b AC = $\sqrt{(-1 - 7)^2 + (3 - 1)^2}$

$= \sqrt{(-8)^2 + 2^2}$

$= \sqrt{68}$

≈ 8.25 units

c BC = $\sqrt{(-1 - -3)^2 + (3 - -5)^2}$

$= \sqrt{2^2 + 8^2}$

$= \sqrt{68}$

$= $ AC

d Gradient of AC = $\dfrac{3 - 1}{-1 - 7} = \dfrac{2}{-8} = -\dfrac{1}{4}$

e gradient of BC $\times$ gradient of AC $= 4 \times -\frac{1}{4}$
$$= -1$$

$\therefore$ AC is perpendicular to BC. The angle at C is $90°$.

f Area $= \frac{1}{2}$ base $\times$ height
$$= \frac{1}{2} \times \sqrt{68} \times \sqrt{68}$$
$$= 34 \text{ units}^2$$

g Gradient of AB $= \dfrac{-5-1}{-3-7} = \dfrac{-6}{-10} = \dfrac{3}{5}$

$\therefore$ the equation of AB is $y = \frac{3}{5}x + c$

Using the point $(7, 1)$, $1 = \frac{3}{5}(7) + c$
$$\therefore \quad 1 = \frac{21}{5} + c$$
$$\therefore \quad 1 - \frac{21}{5} = c$$
$$\therefore \quad c = -\frac{16}{5}$$

So, AB is $y = \frac{3}{5}x - \frac{16}{5}$

which is $5y = 3x - 16$

or $3x - 5y = 16$

h D is the midpoint of AB

$\therefore$ D $= \left(\dfrac{7 + -3}{2}, \dfrac{1 + -5}{2}\right)$ or $(2, -2)$

But DC is perpendicular to AB

$\therefore$ its gradient is $-\frac{5}{3}$.
$$\therefore \quad y = -\frac{5}{3}x + c$$
Using the point $(2, -2)$, $-2 = -\frac{5}{3}(2) + c$
$$\therefore \quad -2 = -\frac{10}{3} + c$$
$$\therefore \quad -2 + \frac{10}{3} = c$$
$$\therefore \quad c = \frac{4}{3}$$
$\therefore$ the equation of CD is $y = -\frac{5}{3}x + \frac{4}{3}$.

2 a There are twelve equal angles at the centre of the dodecagon.
$$\therefore \quad \widehat{\text{AOB}} = \frac{360°}{12} = 30°$$

b

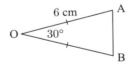

Area of $\triangle$AOB
$$= \frac{1}{2} \times 6 \times 6 \times \sin 30°$$
$$= 9 \text{ cm}^2$$

c Area of dodecagon $= 12 \times 9$
$$= 108 \text{ cm}^2$$

d Area of circle $= \pi r^2$
$$= 36\pi \text{ cm}^2$$

e **i** Percentage error $= \dfrac{108 - 36\pi}{36\pi} \times 100\%$
$$\approx -4.51\%$$

ii Using the cosine rule,
$$\text{AB}^2 = 6^2 + 6^2 - 2 \times 6 \times 6 \times \cos 30°$$
$$\therefore \quad \text{AB} \approx 3.1058 \text{ cm}$$
$$\therefore \quad \text{the perimeter of the dodecagon} \approx 12 \times 3.1058$$
$$\approx 37.270 \text{ cm}$$
The circumference of the circle $= 2\pi r$
$$= 12\pi \text{ cm}$$
$$\therefore \quad \text{the percentage error} \approx \frac{37.270 - 12\pi}{12\pi} \times 100\%$$
$$\approx -1.14\%$$

3 a

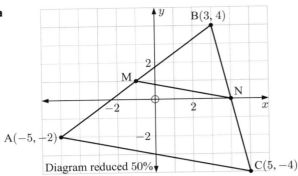

Diagram reduced 50%

b Midpoint of AB is M$\left(\dfrac{-5+3}{2}, \dfrac{-2+4}{2}\right)$ or $(-1, 1)$.

c Midpoint of BC is N$\left(\dfrac{3+5}{2}, \dfrac{4+-4}{2}\right)$ or $(4, 0)$.

d Gradient of MN $= \dfrac{0-1}{4--1} = -\dfrac{1}{5}$

e The equation of the line through M and N is
$$y = -\frac{1}{5}x + c$$
Using the point $(-1, 1)$,
$$\therefore \quad 1 = -\frac{1}{5}(-1) + c$$
$$\therefore \quad 1 = \frac{1}{5} + c \text{ and so } c = \frac{4}{5}$$
So, the line MN is $y = -\frac{1}{5}x + \frac{4}{5}$

which is $5y = -x + 4$

or $x + 5y = 4$

f gradient AC $= \dfrac{-4--2}{5--5} = -\dfrac{1}{5} = $ gradient MN

$\therefore$ AC is parallel to MN.

g Length AC $= \sqrt{(5--5)^2 + (-4--2)^2}$
$$= \sqrt{10^2 + (-2)^2}$$
$$= \sqrt{104}$$
$$\approx 10.2 \text{ cm}$$
and MN $= \sqrt{26} \approx 5.1 \text{ cm}$

$\therefore$ AC is twice the length of MN.

h **i**

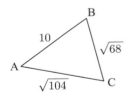

Using the cosine rule,
$$\cos \widehat{\text{ABC}} = \frac{10^2 + \left(\sqrt{68}\right)^2 - \left(\sqrt{104}\right)^2}{2 \times 10 \times \sqrt{68}}$$
$$\approx 0.388$$
$$\therefore \quad \widehat{\text{ABC}} \approx \cos^{-1}(0.388) \approx 67.2°$$

ii Area $\triangle$ABC $= \frac{1}{2} \times 10 \times \sqrt{68} \sin \widehat{\text{ABC}}$
$$= 38 \text{ units}^2$$

iii

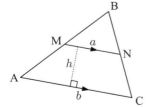

$\triangle$ABC has area 38 units^2 and $\triangle$BMN has area 9.5 units^2.

$\therefore$ trapezium AMNC has area $\quad 38 - 9.5$
$$= 28.5 \text{ units}^2$$

Now area of trapezium $= h\left(\dfrac{a + b}{2}\right)$

$$\therefore \quad 28.5 = h\left(\dfrac{\sqrt{26} + \sqrt{104}}{2}\right)$$

$$\therefore \quad 57 = h\left(\sqrt{26} + \sqrt{104}\right)$$

$$\therefore \quad h = \dfrac{57}{(\sqrt{26} + \sqrt{104})}$$

$$\approx 3.73 \text{ units}$$

$\therefore$ the perpendicular distance is 3.73 units.

4 a M is at $\left(\dfrac{0 + 4}{2}, \dfrac{2 + 0}{2}\right)$ or $(2, 1)$.

b If B has coordinates (x, y)

then $\quad \dfrac{x - 3}{2} = 2 \quad$ and $\quad \dfrac{y - 2}{2} = 1$

$\therefore \quad x - 3 = 4 \quad$ and $\quad y - 2 = 2$

$\therefore \quad x = 7 \quad$ and $\quad y = 4$

So, B is at $(7, 4)$.

c $AD = \sqrt{(-3 - 0)^2 + (-2 - 2)^2}$
$= \sqrt{25} = 5 \text{ units}$

d $AC = \sqrt{(4 - 0)^2 + (0 - 2)^2}$
$= \sqrt{20} \text{ units}$
$CD = \sqrt{(-3 - 4)^2 + (-2 - 0)^2}$
$= \sqrt{53} \text{ units}$

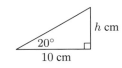

Using the cosine rule,

$$\cos(\widehat{ADC}) = \dfrac{(\sqrt{53})^2 + 5^2 - (\sqrt{20})^2}{2 \times \sqrt{53} \times 5}$$

$$\approx 0.7967$$

$$\therefore \quad \widehat{ADC} \approx \cos^{-1}(0.7967)$$

$$\approx 37.2°$$

e Area of $\triangle ACD = \frac{1}{2} \times 5 \times \sqrt{53} \times \sin \widehat{ADC}$
$= 11 \text{ units}^2$

$\therefore$ area of parallelogram $= 2 \times 11 = 22 \text{ units}^2$

5 a

$\tan 20° = \dfrac{\text{OPP}}{\text{ADJ}} = \dfrac{h}{10}$

$\therefore \quad h = 10 \tan 20°$
≈ 3.6397
≈ 3.640

b Area of cross-section $= \frac{1}{2} \times 10 \times h$
≈ 18.1985
$\approx 18.2 \text{ cm}^2$

c Volume $=$ area of cross-section $\times$ depth
$\therefore \quad 60 \approx 18.2 \times d$
$\therefore \quad d \approx 3.30$

d Let the hypotenuse of the triangle be x cm.

$\cos 20° = \dfrac{\text{ADJ}}{\text{HYP}} = \dfrac{10}{x}$

$\therefore \quad x = \dfrac{10}{\cos 20°}$
≈ 10.642

Total surface area

$\approx 2 \times 18.1985 + (10.642 + 10 + 3.6397) \times \dfrac{60}{18.1985}$

≈ 116.45

$\approx 116 \text{ cm}^2$

e $116.45 \text{ cm}^2 = 0.011\,645 \text{ m}^2$

$\therefore$ the total surface area of 10 000 doorstops is 116.45 m^2

$\therefore$ the total cost is $\quad 116.45 \times 1.2 \approx \$139.74 \approx \$140$

6 a

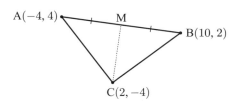

length $AC = \sqrt{(2 - -4)^2 + (-4 - 4)^2}$
$= \sqrt{36 + 64}$
$= \sqrt{100} = 10 \text{ units}$

length $BC = \sqrt{(2 - 10)^2 + (-4 - 2)^2}$
$= \sqrt{64 + 36}$
$= \sqrt{100} = 10 \text{ units}$

b $\triangle ABC$ is isosceles since $\quad AC = BC = 10 \text{ units}$.

c The midpoint of AB is $\left(\dfrac{-4 + 10}{2}, \dfrac{4 + 2}{2}\right)$ which is $(3, 3)$.

d $\widehat{AMC} = 90°$ {altitude bisects base of isosceles $\triangle$}

e i The gradient of $BC = \dfrac{-4 - 2}{2 - 10} = \dfrac{-6}{-8} = \dfrac{3}{4}$

Now gradient of $AC \times$ gradient of BC
$= -\dfrac{4}{3} \times \dfrac{3}{4} = -1$

$\therefore \quad AC$ is perpendicular to BC

$\therefore \quad \widehat{ACB} = 90°$

ii

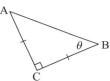

Since $\triangle ABC$ is a right isosceles triangle,
$\widehat{CBA} = 45°$.

f The gradient of $BC = \dfrac{3}{4}$

$\therefore$ the equation of the line through B and C is
$y = \dfrac{3}{4}x + c$.

Using the point $(2, -4)$, $\quad -4 = \dfrac{3}{4}(2) + c$

$\therefore \quad -4 = \dfrac{3}{2} + c$

and so $c = -\dfrac{11}{2}$

So, the line through B and C is
$$y = \dfrac{3}{4}x - \dfrac{11}{2}$$
which is $\quad 4y = 3x - 22$
or $\quad 3x - 4y = 22$

g i The equation of a line parallel to BC is
$3x - 4y = k$ where $k \in \mathbb{R}$.

ii The equation of a line perpendicular to BC is
$4x + 3y = k$ where $k \in \mathbb{R}$.

7

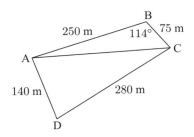

a **i** Using the cosine rule in $\triangle ABC$,
$$AC^2 = 250^2 + 75^2 - 2 \times 250 \times 75 \cos 114°$$
$$\therefore \quad AC \approx 289 \text{ m}$$

ii Using the cosine rule in $\triangle ADC$,
$$\cos A\widehat{D}C \approx \frac{140^2 + 280^2 - 289^2}{2 \times 140 \times 280} \approx 0.1865$$
$$\therefore \quad A\widehat{D}C \approx \cos^{-1}(0.1865) \approx 79.3°$$

iii Area $\triangle ADC = \frac{1}{2} \times 140 \times 280 \sin 79.25°$
$$= 19\,256.1 \approx 19\,300 \text{ m}^2$$

iv Area $\triangle ABC = \frac{1}{2} \times 250 \times 75 \sin 114°$
$$\approx 8564.5 \text{ m}^2$$
$$\therefore \quad \text{total area of park} \approx 19\,256.1 + 8564.5$$
$$\approx 27\,820.6 \approx 27\,800 \text{ m}^2$$

b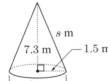

i Let the slant height be s m.
$$\therefore \quad 7.3^2 + 1.5^2 = s^2 \quad \{\text{Pythagoras}\}$$
$$\therefore \quad s^2 = 53.29 + 2.25$$
$$\therefore \quad s = \sqrt{55.54} \approx 7.45$$
So, the slant height is 7.45 m.

ii Let the angle between the base and the slant be θ.
$$\therefore \quad \tan\theta = \frac{7.3}{1.5}$$
$$\therefore \quad \theta = \tan^{-1}\left(\frac{7.3}{1.5}\right)$$
$$\approx 78.4°$$

iii Volume $= \frac{1}{3}\pi r^2 h$
$$= \frac{1}{3}\pi(1.5)^2 \times 7.3$$
$$\approx 17.2 \text{ m}^3$$
$$\therefore \quad \text{total volume} \approx 34.4 \text{ m}^3$$

iv Area of curved surface $= \pi r s$
$$\approx \pi \times 1.5 \times 7.45$$
$$\approx 35.1 \text{ m}^2$$
$$\therefore \quad \text{total area of curved surface is } 70.2 \text{ m}^2.$$
We require $70.2 \div 15 \approx 4.68$ litres of paint.

8 **a** **i**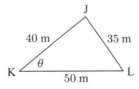

ii Let $J\widehat{K}L = \theta$
Using the cosine rule,
$$\cos\theta = \frac{40^2 + 50^2 - 35^2}{2 \times 40 \times 50} \approx 0.719$$
$$\theta \approx \cos^{-1}(0.719) \approx 44.0°$$
So, Kim must turn through $44.0°$.

b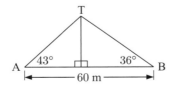

i $A\widehat{T}B = 180° - (43° + 36°) \quad \{\text{angle sum of } \triangle\}$
$$= 101°$$

ii $\dfrac{AT}{\sin 36°} = \dfrac{60}{\sin 101°} \quad \{\text{sine rule}\}$
$$\therefore \quad AT = \frac{60 \sin 36°}{\sin 101°} \approx 35.9$$
So, the treasure is 35.9 m from point A.

iii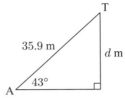

Let the vertical distance be d m.
Now $\sin 43° \approx \dfrac{d}{35.9}$
$$\therefore \quad 35.9 \sin 43° \approx d$$
$$\therefore \quad 24.5 \approx d$$
So, the vertical distance is 24.5 m.

iv Area of $\triangle ATB \approx \frac{1}{2} \times 35.9 \times 60 \sin 43°$
$$\approx 735 \text{ m}^2$$

9 **a**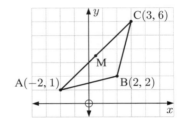

b Length $AC = \sqrt{(3-(-2))^2 + (6-1)^2}$
$$= \sqrt{50}$$
$$\approx 7.07 \text{ units}$$

c Midpoint M is at $\left(\dfrac{-2+3}{2}, \dfrac{1+6}{2}\right)$ or $\left(\frac{1}{2}, \frac{7}{2}\right)$

d $BM = \sqrt{\left(2-\frac{1}{2}\right)^2 + \left(2-\frac{7}{2}\right)^2} = \sqrt{4.5}$ units
$BC = \sqrt{(3-2)^2 + (6-2)^2} = \sqrt{17}$ units
$CM = \frac{1}{2} \times \sqrt{50} = \sqrt{12.5}$ units
Using the cosine rule,
$$\cos M\widehat{B}C = \frac{4.5 + 17 - 12.5}{2 \times \sqrt{4.5} \times \sqrt{17}}$$
$$\therefore \quad M\widehat{B}C \approx 59.0°$$

e Gradient of $AC = \dfrac{6-1}{3-(-2)} = 1$

f Gradient of $BM = \dfrac{\frac{7}{2}-2}{\frac{1}{2}-2} = \dfrac{\frac{3}{2}}{-\frac{3}{2}} = -1$

g Using **f**, the equation has the form $y = -x + c$.
Using the point $B(2, 2)$, $2 = -2 + c$
$$\therefore \quad c = 4$$
$\therefore$ the equation of BM is $y = -x + 4$.

10 **a** Using the cosine rule,
$$(BC)^2 = 65^2 + 104^2 - 2 \times 65 \times 104 \times \cos 60°$$
$$= 8281$$
$$\therefore \quad BC = 91 \text{ m}$$

b Area $= \frac{1}{2} \times 65 \times 104 \times \sin 60°$

$\qquad \approx 2927.2$

$\qquad \approx 2930 \text{ m}^2$

c **i** Area $A_1 = \frac{1}{2} \times 65 \times x \times \sin 30°$

$\qquad = \frac{65}{4}x \text{ m}^2$

ii Area $A_2 = \frac{1}{2} \times 104 \times x \times \sin 30°$

$\qquad = 26x \text{ m}^2$

iii $\qquad A_1 + A_2 = \text{total area}$

$\qquad \therefore \quad \frac{65}{4}x + 26x \approx 2927.2$

$\qquad \therefore \quad \frac{169}{4}x \approx 2927.2$

$\qquad \therefore \quad x \approx 69.28$

$\qquad \therefore \quad x \approx 69.3$

d Length $\text{BD} = \frac{5}{8} \times \text{length DC}$

$\qquad \therefore \quad \text{BD} = \frac{5}{13} \times \text{length of BC}$

$\qquad = 35 \text{ m} \quad \{\text{using } \mathbf{a}\}$

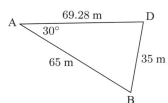

Using the sine rule,

$$\frac{\sin \widehat{\text{ADB}}}{65} = \frac{\sin 30°}{35}$$

$$\therefore \quad \sin \widehat{\text{ADB}} = \frac{65 \sin 30°}{35}$$

$$\therefore \quad \widehat{\text{ADB}} \approx 68.2°$$

11 **a**

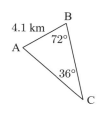

$$\frac{\text{AC}}{\sin 72°} = \frac{4.1}{\sin 36°} \quad \{\text{sine rule}\}$$

$$\therefore \quad \text{AC} = \frac{4.1 \sin 72°}{\sin 36°}$$

$$\approx 6.63 \text{ km}$$

Since EF is 20% longer than AC,

$\qquad$ length $\text{EF} = 1.2 \times \text{length AC}$

$\qquad\qquad \approx 7.96 \text{ km}$

b

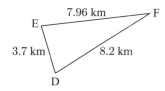

Using the cosine rule,

$$\cos \widehat{\text{DEF}} \approx \frac{7.96^2 + 3.7^2 - 8.2^2}{2 \times 3.7 \times 7.96} \approx 0.167$$

$$\widehat{\text{DEF}} \approx 80.4°$$

c

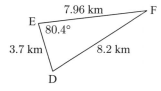

Area $\triangle \text{DEF} \approx \frac{1}{2} \times 3.7 \times 7.96 \sin 80.4°$

$\qquad\qquad \approx 14.5 \text{ km}^2$

d $\widehat{\text{BAC}} = 180° - (72° + 36°) = 72°$

$\qquad \therefore \quad \triangle \text{ABC is isosceles}$

$\qquad \therefore \quad \text{BC} = \text{AC} \approx 6.63 \text{ km}$

$\qquad \therefore \quad$ length of course 1 $\approx 4.1 + 6.63 + 6.63$

$\qquad\qquad\qquad \approx 17.4 \text{ km}$

e Course 2 has length $\approx 3.7 + 8.2 + 7.96 \approx 19.9 \text{ km}$

$\qquad$ Time to complete course 1 $\approx \frac{17.4}{14} \text{ h}$

$\qquad\qquad \approx 74.4 \text{ min}$

$\qquad\qquad \approx 1 \text{ hour } 14 \text{ minutes}$

$\qquad$ Time to complete course 2 $\approx \frac{19.9}{10} \text{ h}$

$\qquad\qquad = 1.99 \text{ hours}$

$\qquad\qquad = 1 \text{ hour } 59 \text{ minutes}$

$\qquad \therefore \quad$ it will take Wael 45 minutes extra to complete course 2 (to the nearest 5 minutes).

12

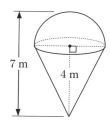

a Height of the hemisphere is $7 - 4 = 3 \text{ m}$

$\qquad \therefore \quad$ radius of cone $= 3 \text{ m}$

b Volume $= \frac{1}{3}\pi r^2 h + \frac{2}{3}\pi r^3$

$\qquad = \frac{1}{3}\pi \times 3^2 \times 4 + \frac{2}{3}\pi \times 3^3$

$\qquad \approx 94.2 \text{ m}^3$

c

$\tan \theta = \frac{4}{3}$

$\theta = \tan^{-1}\left(\frac{4}{3}\right)$

$\approx 53.1°$

d slant height $\quad s = \sqrt{3^2 + 4^2} \quad \{\text{Pythagoras}\}$

$\qquad\qquad\qquad = \sqrt{25}$

$\qquad\qquad\qquad = 5 \text{ m}$

e $\quad$ Total surface area

$\qquad = \text{surface area of hemisphere} + \text{surface area of cone}$

$\qquad = \frac{1}{2} \times 4\pi r^2 + \pi r s$

$\qquad = \frac{1}{2} \times 4 \times \pi \times 3^2 + \pi \times 3 \times 5$

$\qquad \approx 104 \text{ m}^2$

f Weight of ice cream $\approx 104 \times 1.23$

$\qquad\qquad\qquad \approx 128 \text{ kg}$

13 **a**

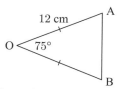

Using the cosine rule,

$\qquad (\text{AB})^2 = 12^2 + 12^2 - 2 \times 12 \times 12 \times \cos 75°$

$\qquad \therefore \quad \text{AB} \approx 14.6 \text{ cm}$

b Area of $\triangle \text{OAB} = \frac{1}{2} \times 12 \times 12 \times \sin 75°$

$\qquad\qquad\qquad \approx 69.5 \text{ cm}^2$

c $\triangle \text{OAB}$ is isosceles, so $\widehat{\text{OAB}} = \left(\frac{180 - 75}{2}\right)° = 52\frac{1}{2}°$

$\qquad$ Since AP is a tangent, $\widehat{\text{BAP}} = 90° - 52\frac{1}{2}° = 37\frac{1}{2}°$

$\qquad$ Also, $\widehat{\text{APB}} = 105° \quad \{\text{angle sum of quadrilateral}\}$

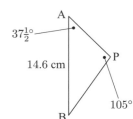

Using the sine rule,

$$\frac{BP}{\sin 37\frac{1}{2}^\circ} \approx \frac{14.6}{\sin 105^\circ}$$

$$\therefore \quad BP \approx \frac{14.6 \sin 37\frac{1}{2}^\circ}{\sin 105^\circ}$$

$$\therefore \quad BP \approx 9.21 \text{ cm}$$

d By the same reasoning as in **c**, AP ≈ 9.21 cm

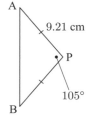

Area of △ABP
$$\approx \frac{1}{2} \times 9.21 \times 9.21 \times \sin 105^\circ$$
$$\approx 40.9 \text{ cm}^2$$

14 a There are 6 equal angles in the centre of the hexagon.

$$\therefore \quad A\widehat{O}B = \frac{360^\circ}{6} = 60^\circ$$

b OA = OB, so

$$O\widehat{A}B = O\widehat{B}A = \left(\frac{180 - 60}{2}\right)^\circ = 60^\circ$$

All the angles are 60°, so △OAB is equilateral.

c The sign is made up of 6 equilateral triangles.
The area of each triangle
$$= \frac{1}{2} \times 20 \times 20 \times \sin 60^\circ$$
$$\approx 173.2 \text{ cm}^2$$

$$\therefore \quad \text{total area of figure} = 6 \times 173.2$$
$$= 1039.2$$
$$\approx 1040 \text{ cm}^2$$

d i The height of the sign y = length OA × 2
$$= 40 \text{ cm}$$

ii Now $\tan 30^\circ = \dfrac{10}{d}$

$$\therefore \quad d = \frac{10}{\tan 30^\circ}$$
$$\approx 17.32$$

The width of the sign $x \approx 2 \times 17.32 \approx 34.6$ cm

iii Area = height × width
$$\approx 40 \times 34.6$$
$$\approx 1385.6 \text{ cm}^2$$
$$\approx 1390 \text{ cm}^2$$

iv Wasted area = area of rectangle − area of hexagon
$$\approx 1385.6 - 1039.2$$
$$\approx 346 \text{ cm}^2$$

$$\therefore \quad \text{proportion wasted} \approx \frac{346}{1386} \times 100\%$$
$$= 25\%$$

v The wasted area ≈ 346 cm²
$$\approx 0.0346 \text{ m}^2$$

$$\therefore \quad \text{the cost of the wasted material}$$
$$\approx \text{€}350 \times 0.0346$$
$$\approx \text{€}12.11$$

SOLUTIONS TO TOPIC 6 (MATHEMATICAL MODELS)

SHORT QUESTIONS

1 a gradient $m = \dfrac{y\text{-step}}{x\text{-step}} = \dfrac{2-1}{2-0} = \dfrac{1}{2}$

y-intercept = 1

$$\therefore \quad C(t) = \frac{1}{2}t + 1$$

b $C(23) = \frac{1}{2}(23) + 1 = \$12.50$

$\therefore$ the call costs \$12.50.

c If $C(t) = \$18.31$

then $\frac{1}{2}t + 1 = 18.31$

$$\therefore \quad \frac{1}{2}t = 17.31$$
$$\therefore \quad t = 34.62$$

$\therefore$ the call lasts about 35 minutes.

2 a $h(t) = 1 - 2t^2$

$$\therefore \quad h(0) = 1 - 2(0)^2$$
$$= 1$$

b If $h(t) = 0$

then $1 - 2t^2 = 0$

$$\therefore \quad 2t^2 = 1$$
$$\therefore \quad t^2 = \frac{1}{2}$$
$$\therefore \quad t = \pm\frac{1}{\sqrt{2}}$$

c Since $h(t)$ is a quadratic, its domain = $\{t \mid t \in \mathbb{R}\}$.

d

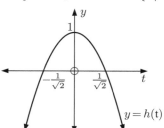

From the graph of $h(t)$, we observe the maximum is at $(0, 1)$.

$\therefore$ the range of $h(t)$ is $\{y \mid y \leqslant 1, \ y \in \mathbb{R}\}$.

3 a i The percentage of Carbon-14 remaining after 4 thousand years ≈ 61%.

ii It will take about 5600 years for the percentage of Carbon-14 to fall to 50%.

b When $t = 19$, $P = 100 \times (1.1318)^{-19}$
$$\approx 9.51$$

After 19 thousand years there will be ≈ 9.51% remaining.

c The asymptote has equation $P = 0$.

4 a $f(2) = 15 - 2(2) = 11$ **b** $g(-2) = 2^{-2} + 1 = 1\frac{1}{4}$

c $g(x) = f(x)$ when $2^x + 1 = 15 - 2x$

Using technology, $x = 3$

5 a We see that for every increase of 5 bags of rice, the price decreases by 2000 rupiah.

$\therefore$ $P(b)$ is a linear function with gradient $\frac{-2000}{5} = -400$

$\therefore$ $P(b) = -400b + c$ for some constant c.

Now $P(50) = 30\,000$

$$\therefore \quad -400 \times 50 + c = 30\,000$$
$$\therefore \quad c = 50\,000$$
$$\therefore \quad P(b) = -400b + 50\,000$$

b $P(60) = -400 \times 60 + 50\,000 = 26\,000$

∴ the total cost $= 60 \times 26\,000 = 1\,560\,000$ rupiah

6 a The x-intercepts are at 2 and -1, so $y = a(x-2)(x+1)$

The graph passes through $(3, 12)$

$$\therefore \quad a(1)(4) = 12$$
$$\therefore \quad a = 3$$
$$\text{So,} \quad y = 3(x-2)(x+1)$$
$$= 3(x^2 - x - 2)$$
$$= 3x^2 - 3x - 6$$

b The axis of symmetry is $x = -\dfrac{b}{2a} = \dfrac{3}{6} = \dfrac{1}{2}$

Now when $x = \frac{1}{2}$, $\quad y = 3(\frac{1}{2})^2 - 3(\frac{1}{2}) - 6$
$$= \frac{3}{4} - \frac{3}{2} - 6 = -\frac{27}{4}$$

∴ the vertex is at $(\frac{1}{2}, -\frac{27}{4})$.

7 a The graph passes through $(0, 20)$ and $(2, 35)$.

$$\therefore \quad a + b = 20 \;\; .... \;(1)$$
$$\text{and} \quad 4a + b = 35 \;\; .... \;(2)$$

b Subtracting (1) from (2), $\quad 3a = 15$
$$\therefore \quad a = 5$$
$$\text{Using (1),} \quad b = 15$$

c $y = 5 \times 2^x + 15$

When $x = 1$, $\quad y = 5 \times 2 + 15 = 25$
$$\therefore \quad p = 25$$

When $x = 3$, $\quad y = 5 \times 2^3 + 15 = 55$
$$\therefore \quad q = 55$$

8 a Domain $= \{0, 1, 2, 3\}$ **b** Range $= \{-4, -2, 0, 2\}$

c

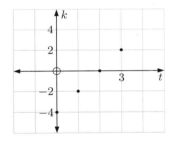

9 a $S_n = \dfrac{n}{2}(2a + (n-1)d)$

$$\therefore \quad \frac{n}{2}(2 \times 7 + (n-1)(-5)) = -1001$$
$$\therefore \quad n(14 - 5n + 5) = -2002$$
$$\therefore \quad 5n^2 - 19n - 2002 = 0$$

b Using technology, $n = -\frac{91}{5}$ or 22

But $n \in \mathbb{Z}^+$, so $n = 22$

10 a $N(0) = 120 \times (1.04)^0 = 120$

∴ the settlement started with 120 people.

b $N(4) = 120(1.04)^4 \approx 140.4$

∴ there were 140 people after 4 years.

c If $N(t) = 240$

then $120 \times (1.04)^t = 240$
$$\therefore \quad (1.04)^t = \frac{240}{120} = 2$$
$$\therefore \quad t \approx 17.7$$

∴ it will take 18 years for the population to double.

11 a i $f(-4) = \sqrt{0} = 0$ **ii** $f(0) = \sqrt{4} = 2$
iii $f(12) = \sqrt{16} = 4$

b

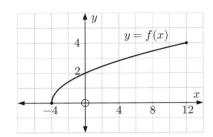

c Range $= \{y \mid 0 \leqslant y \leqslant 4, \; y \in \mathbb{R}\}$

12 a When $d = 2$, $\quad N = 810$

$$\therefore \quad k(2)^{-3} = 810$$
$$\therefore \quad \frac{k}{8} = 810$$
$$\therefore \quad k = 6480$$

b $N = k(3)^{-3} = \frac{6480}{27} = 240$

So, 240 ball bearings can be made.

c If $N = 23$, $\quad kd^{-3} = 23$
$$\therefore \quad \frac{6480}{d^3} = 23$$
$$\therefore \quad d^3 = \frac{6480}{23}$$
$$\therefore \quad d \approx 6.56$$
$$\therefore \quad r \approx \frac{6.56}{2} \approx 3.28 \text{ mm}$$

13 a

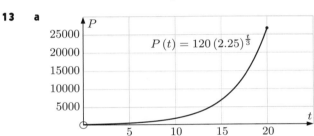

b $P(10) = 120(2.25)^{\frac{10}{3}} \approx 1790$ bees

c When $P = 5000$, $\quad 120(2.25)^{\frac{t}{3}} = 5000$

Using technology, $\quad t \approx 13.8$

So, it will take about 13.8 weeks.

14 a The y-intercept $= 9$, so $c = 9$

b The axis of symmetry is $\quad x = -\dfrac{b}{2a}$

$$\therefore \quad -\frac{b}{2a} = 1$$
$$\therefore \quad -b = 2a$$
$$\therefore \quad 2a + b = 0 \;\; .... \;(1)$$

c The point $(1, 7)$ lies on the graph,

so $a(1)^2 + b(1) + 9 = 7$
$$\therefore \quad a + b = -2 \;\; .... \;(2)$$

d Solving (1) and (2) simultaneously, $a = 2$ and $b = -4$.

15 a Graphing $y = x^3 - 3x^2 - x + 3$ on $-2 \leqslant x \leqslant 3$:

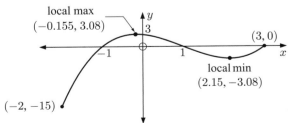

b Range $= \{y \mid -15 \leqslant y \leqslant 3.08, \; y \in \mathbb{R}\}$

16 a $H(0) = 0$ m

The object begins at ground level.

b $H(t) = 19.6t - 4.9t^2$
$\quad\quad = 4.9t(4 - t)$

$\therefore\quad H(t) = 0$ when $t = 0$ and $t = 4$

The object returns to ground level after 4 seconds.

c Domain of $H(t) = \{t \mid 0 \leqslant t \leqslant 4, \ t \in \mathbb{R}\}$

d The maximum height occurs when

$$t = -\frac{b}{2a} = \frac{-19.6}{-9.8} = 2 \text{ seconds.}$$

$H(2) = 19.6$, so the maximum height is 19.6 m.

17 a $(x + 7)(x - 4) = x^2 + 7x - 4x - 28$
$\quad\quad\quad\quad\quad\quad\quad = x^2 + 3x - 28$

b Using **a**, the zeros of $x^2 + 3x - 28$ are -7 and 4.
We hence find $A(-7, 0)$ and $B(4, 0)$.

c The axis of symmetry is midway between the x-intercepts.

$\therefore\quad$ its equation is $x = \dfrac{-7 + 4}{2} = -\dfrac{3}{2}$.

d When $x = -\frac{3}{2}$, $y = \left(-\frac{3}{2}\right)^2 + 3\left(-\frac{3}{2}\right) - 28 = -30\frac{1}{4}$

So, C is $\left(-1\frac{1}{2}, -30\frac{1}{4}\right)$.

18 a $f(x) = 8x - 2x^2 = 2x(4 - x)$

b The x-intercepts are 0 and 4.

c The axis of symmetry is midway between the x-intercepts.
$\therefore\quad$ its equation is $x = 2$.

d $f(2) = 8(2) - 2(2)^2 = 8$
$\therefore\quad$ the vertex is $(2, 8)$.

19 We use technology to sketch the function.

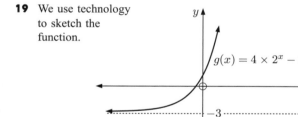

The range is $\{y \mid y > -3, \ y \in \mathbb{R}\}$.

20 a $N(0) = 30 - 3^3 = 3$

So, there were initially 3000 ants in the colony.

b $N(2) = 30 - 3^{3-2} = 27$

So, there were 27 000 ants after two months.

c $N(t) = 20$ when $30 - 3^{3-t} = 20$
$\quad\quad\quad\quad\quad\quad\quad \therefore\ 3^{3-t} = 10$
$\quad\quad\quad\quad\quad\quad\quad\quad\quad \therefore\ t \approx 0.904$ months

So, the colony will reach a population of 20 000 in the first month.

d The horizontal asymptote is $N = 30$.

e The population should never reach the asymptote, which is at 30 000 ants.

21 a $f(0) = \dfrac{2^0}{0 - 1} = -1$

$\therefore\quad$ the y-intercept is -1

b Using technology, the minimum for $x > 1$ is at $(2.44, 3.77)$. So, the minimum value is 3.77.

c The vertical asymptote is $x = 1$.

d $f(5) = \dfrac{2^5}{5 - 1} = \frac{32}{4} = 8$

e Graph of $y = \dfrac{2^x}{x - 1}$

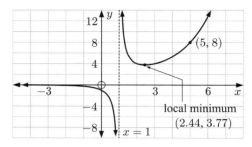

22 a The axis of symmetry has equation $x = -\dfrac{m}{2}$.

This axis passes through the vertex, so $-\dfrac{m}{2} = 1$
$$\therefore\ m = -2$$

Since $V(1, 3)$ lies on the quadratic, $1 - 2 + n = 3$
$$\therefore\ n = 4$$

b $\quad f(x) = x^2 - 2x + 4$
$\therefore\ f(3) = 9 - 6 + 4 = 7$
$\quad\quad \therefore\ k = 7$

c Domain $\{x \mid x \in \mathbb{R}\}$, Range $\{y \mid y \geqslant 3, \ y \in \mathbb{R}\}$

23 a Graph of $y = 3 + \dfrac{1}{x - 2}$

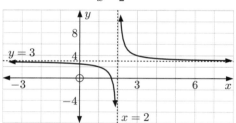

b Vertical asymptote is $x = 2$
Horizontal asymptote is $y = 3$

24 a $\quad\quad\quad\quad T(t) = A \times B^{-t} + 3$
$\quad\quad\quad\quad T(0) = 27$
$\therefore\ A \times B^0 + 3 = 27$
$\quad\quad \therefore\ A = 24$

b $\quad\quad\quad\quad T(3) = 6$
$\therefore\ 24 \times B^{-3} + 3 = 6$
$\quad\quad\quad \therefore\ \dfrac{24}{B^3} = 3$
$\quad\quad\quad\quad \therefore\ B^3 = 8$ and so $B = 2$

c $\quad T(5) = 24 \times 2^{-5} + 3$
$\therefore\ T(5) = 3.75\,°C$

d $T(t) = 24 \times 2^{-t} + 3$ has an asymptote at $T = 3\,°C$.
The temperature will approach 3°C, but will never reach it.

25

Graph	Function
a	E
b	A
c	C
d	D

26 **a** $f(x) = ax^2 + bx + 7$

$f(2) = 7$, then $4a + 2b + 7 = 7$

$\therefore \quad 4a + 2b = 0$

$2a + b = 0$ .... (1)

$f(4) = 23$, then $16a + 4b + 7 = 23$

$\therefore \quad 16a + 4b = 16$

$\therefore \quad 4a + b = 4$ .... (2)

b Subtracting (1) from (2), $\quad 2a = 4$

$\therefore \quad a = 2$

Using (1), $\quad b = -4$.

c $\quad\quad f(x) = 2x^2 - 4x + 7$

$\therefore \quad f(-1) = 2 + 4 + 7$

$= 13$

27 **a** Graph of $P(x) = -50x^2 + 1000x - 2000$

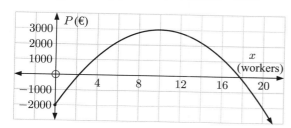

b 10 workers maximise the profit.

c The maximum profit is €3000.

28

Function	Graph
a	E
b	D
c	B

29 **a** **i** The x-intercepts are 0, 2, and 3.

ii local maximum $(1.37, 2.64)$,

local minimum $(2.63, -4.24)$

b

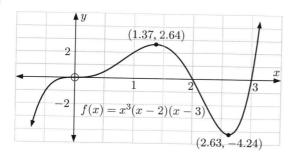

30 **a** Since $y = a(x - 1)(x - 5)^2$, the graph cuts the x-axis at $x = 1$, and touches it at $x = 5$.

$\therefore \quad b = 1$ and $c = 5$

b $y = a(x - 1)(x - 5)^2$

When $x = 0$, $y = -50$

$\therefore \quad a(-1)(-5)^2 = -50$

$\therefore \quad -25a = -50$

$\therefore \quad a = 2$

c Using technology, the local maximum P is $(2.33, 19.0)$.

31 **a** Graph of $y = x^4 - 2x^3 - 3x^2 + 8x - 4$

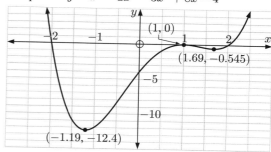

b Local minima are $(1.69, -0.545)$ and $(-1.19, -12.4)$. Local maximum is $(1, 0)$.

c $x^4 - 2x^3 - 3x^2 + 8x - 4 = 0$

when $x = 1$ or $x = \pm 2$

32 **a** $f(2) = 3 - 4^{-2} \approx 2.94$

$\therefore \quad p \approx 2.94$

$f(-2) = 3 - 4^2 = -13$

$\therefore \quad q = -13$

b **i** Using technology, the x-intercept ≈ -0.792.

$f(0) = 3 - 4^0 = 2$ $\therefore$ the y-intercept is 2.

ii As $x \to \infty$, $4^{-x} \to 0$ and so $y \to 3$

$\therefore \quad y = 3$ is the horizontal asymptote.

c

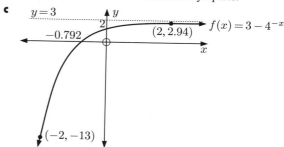

d Range $= \{y \mid y < 3, \ y \in \mathbb{R}\}$

33 **a** $h(-2) = (-2)^2 - 2^{-(-2)} + \dfrac{1}{-2}$

$= 4 - 4 - \frac{1}{2} = -\frac{1}{2}$

b Using technology, $h(x) = 2$ when $x \approx 0.381$ or 1.28.

c The vertical asymptote is $x = 0$.

d

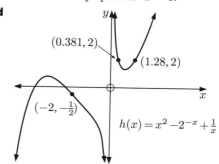

e Using technology, the range is $\{y \mid y < 0.741$ or $y > 1.30, \ y \in \mathbb{R}\}$.

LONG QUESTIONS

1 **a** **i** $C(20) = 20^2 + 400 = 800$

The weekly cost for producing 20 DVD players is $800.

ii $I(20) = 50(20) = 1000$

The weekly income when 20 DVD players are sold is $1000.

iii $P(20) = I(20) - C(20)$
$$= 1000 - 800$$
$$= 200$$

So, a profit of \$200 is made.

b $P(x) = I(x) - C(x)$
$\therefore\ P(x) = 50x - (x^2 + 400)$ dollars

c $P(x) = -x^2 + 50x - 400$

The vertex occurs when $x = -\dfrac{b}{2a} = \dfrac{-50}{2(-1)} = 25$

Hence, 25 DVD players must be made and sold to maximise the profit.

d $P(25) = -(25)^2 + 50(25) - 400 = \$225$

Profit **per** DVD player $= \dfrac{\$225}{25} = \$9$

e The function $P(x)$ has zeros at $x = 10$ and $x = 40$. In order to make positive (non-zero) profit, at most 39 DVD players can be made.

2 a x-intercepts occur when $y = 0$
$$\therefore\ 2 + \frac{4}{x + 1} = 0$$
$$\therefore\ \frac{4}{x + 1} = -2$$
$$\therefore\ 4 = -2(x + 1)$$
$$\therefore\ 2x = -6$$
$$\therefore\ x = -3$$

b y-intercept occurs when $x = 0$
$$\therefore\ y = 2 + \tfrac{4}{1} = 6$$

c $f(-2) = 2 + \dfrac{4}{-2 + 1} = -2$

d i $y = 2$ **ii** $x = -1$

e

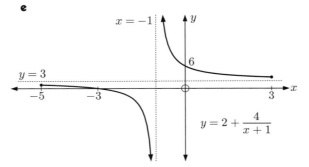

3 a $2x + 2w = 160$
$$\therefore\ 2w = 160 - 2x$$
$$\therefore\ w = 80 - x$$

The width is $(80 - x)$ m.

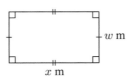

b Area $A(x) = x(80 - x)$ m^2

c

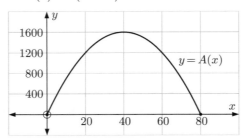

d The maximum area occurs when $x = 40$, which is when the field is a 40 m $\times$ 40 m square.

e i
$$A(x) = 1200$$
$$\therefore\ x(80 - x) = 1200$$
$$\therefore\ x^2 - 80x + 1200 = 0$$
$$\therefore\ (x - 60)(x - 20) = 0$$
$$\therefore\ x = 60 \text{ or } 20$$
$$\therefore\ \text{the field is } 60 \text{ m} \times 20 \text{ m.}$$

ii We lose $1600 - 1200 = 400$ m^2 of productive land
$\therefore$ the lost production $= 400 \times 6.5$
$$= 2600 \text{ kg}$$

4 a
$$N = 1200 \times k^t$$
$$\therefore\ 1200 \times k^4 = 4800$$
$$\therefore\ k^4 = 4$$
$$\therefore\ k = \sqrt{2} \approx 1.41$$

b

t	0	1	2	3	4	5	6
N	1200	1700	2400	3390	4800	6790	9600

c

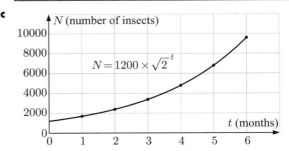

d After $2\tfrac{1}{2}$ months there are about 2900 insects.

e $N = 20\,000$ when $1200 \times \left(\sqrt{2}\right)^t = 20\,000$
$$\therefore\ \left(\sqrt{2}\right)^t = \frac{20\,000}{1200} = \frac{50}{3}$$

Using technology, $t \approx 8.12$

So, it takes about 8.12 months for the population to reach 20 000 insects.

f The percentage change $= \dfrac{N(6) - N(5)}{N(5)} \times 100\%$
$$= \frac{9600 - 6790}{6790} \times 100\%$$
$$\approx 41.4\%$$

5 a i $x = \tfrac{3}{2}$ **ii** $y = 0$

b i When $x = 0$, $y = a^0 = 1$
$\therefore$ the y-intercept is 1.

ii The horizontal asymptote is $y = 0$.

c At $x = 2$, $a^2 = \dfrac{4}{2(2) - 3}$
$$\therefore\ a^2 = 4$$
$$\therefore\ a = \pm 2$$
$$\therefore\ a = 2 \quad \{\text{as } a > 0\}$$

d Graphs of $y = \dfrac{4}{2x - 3}$ and $y = 2^x$:

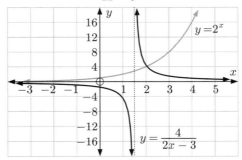

6 a $f(x)$ is undefined when $x = 0$.

b Horizontal asymptote $y = -1$, vertical asymptote $x = 0$.

c $f(1) = \frac{5}{1} - 1 = 4$ $\therefore$ $m = 4$

When $f(x) = 0$, $\frac{5}{x} - 1 = 0$

$\therefore$ $\frac{5}{x} = 1$

$\therefore$ $x = 5$

So, $n = 5$

d, f ii

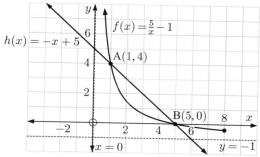

e i
$$\frac{5}{x} - 1 = 5 - x$$
$$\therefore \quad x - 6 + \frac{5}{x} = 0$$
$$\therefore \quad x^2 - 6x + 5 = 0$$
$$\therefore \quad (x - 1)(x - 5) = 0$$

ii Using the Null factor law, $x = 1$ or 5.

f i $y = h(x)$ passes through $(1, 4)$ and $(5, 0)$.

Its gradient $= \frac{0 - 4}{5 - 1} = -1$, so $c = -1$

$\therefore$ $h(x) = -x + d$

Now $h(1) = -1 + d = 4$

$\therefore$ $d = 5$

$\therefore$ $h(x) = -x + 5$

g From the graph, $h(x) \geqslant f(x)$ for positive x when $1 \leqslant x \leqslant 5$.

h The graph passes through the origin, so the y-intercept $r = 0$.

The graph passes through $(1, 4)$, so $p + q = 4$ .... (1)

The graph passes through $(5, 0)$, so

$$25p + 5q = 0 \quad .... (2)$$

Solving (1) and (2) simultaneously, $p = -1$ and $q = 5$

So, $g(x) = -x^2 + 5x$.

7 a, d

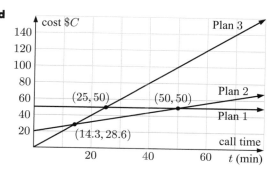

b i *Plan 1*: $50

Plan 2: $20 + 30 \times \$0.60 = \$38$

Plan 3: $30 \times \$2 = \$60$

ii *Plan 1*: $50

Plan 2: $20 + 60 \times \$0.60 = \$56$

Plan 3: $60 \times \$2 = \$120$

c i $C = 50$ **ii** $C = 0.6t + 20$ **iii** $C = 2t$

e i more than 50 **ii** between 14.3 and 50

iii less than 14.3

8 a

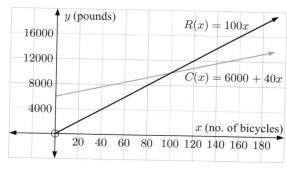

b The initial setup cost $C(0) = \$6000$.

c Revenue $=$ cost when $6000 + 40x = 100x$

$\therefore$ $6000 = 60x$

$\therefore$ $x = 100$

The company breaks even when 100 bicycles are made.

d Total revenue from selling x bicycles $= \$100x$

$\therefore$ the revenue per bicycle $= \$100$

e $P(x) = R(x) - C(x)$

$= 100x - (6000 + 40x)$

$= 60x - 6000$ dollars

f The profit from the sale of 400 bicycles is

$P(400) = 60(400) - 6000 = \$18\,000$

9 a i y-intercept $= f(0) = 0$

ii The maximum is at $(4.33, 4.04)$, so the maximum value of $f(x)$ is 4.04.

iii $f(2) = 2$, $f(-1) = -2$

b, c

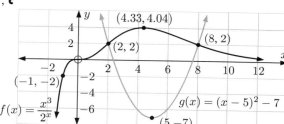

d If $\dfrac{x^3}{2^x} + 7 = (x - 5)^2$

then $\dfrac{x^3}{2^x} = (x - 5)^2 - 7$

$\therefore$ $f(x) = g(x)$

The graphs intersect at $(2, 2)$ and $(8, 2)$, so the solutions are $x = 2$ or 8.

10 a $V = l \times w \times h$

Given that $v = 2000$ cm^3 and $l = 2w$

$\therefore$ $2000 = 2w \times w \times h$

$\therefore$ $\dfrac{2000}{2w^2} = h$

$\therefore$ $h = \dfrac{1000}{w^2}$

b $A = 2lw + 2lh + 2wh$

$= 2 \times 2w \times w + 2 \times 2w \times \dfrac{1000}{w^2} + 2w \times \dfrac{1000}{w^2}$

$= 4w^2 + \dfrac{4000}{w} + \dfrac{2000}{w}$

$= 4w^2 + \dfrac{6000}{w}$ cm^2

c w is a length, so it must be non-negative.

Hence, the domain is $D = \{w \mid w > 0, \ w \in \mathbb{R}\}$.

d

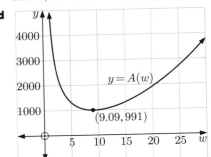

e **i** The minimum surface area is about 991 cm^2.

ii This occurs when $w \approx 9.09$ cm

$\therefore \ l \approx 18.2$ cm and $h \approx 12.1$ cm

$\therefore$ the dimensions

$l \times w \times h \approx 18.2$ cm $\times 9.09$ cm $\times 12.1$ cm.

f The range is $\{A \mid A > 991, \ A \in \mathbb{R}\}$.

11 a

Graph	Domain
A	$\{x \mid x \geqslant -2, \ x \in \mathbb{R}\}$
B	$\{x \mid x \in \mathbb{R}\}$
C	$\{x \mid x \geqslant -4, \ x \in \mathbb{R}\}$
D	$\{x \mid x \in \mathbb{R}, \ x \neq 0\}$

The domains of all of the graphs are different.

b

Graph	Range
A	$\{y \mid y \leqslant 5, \ y \in \mathbb{R}\}$
B	$\{y \mid y > -2, \ y \in \mathbb{R}\}$
C	$\{y \mid y \geqslant 0, \ y \in \mathbb{R}\}$
D	$\{y \mid y \neq 0, \ y \in \mathbb{R}\}$

c

Graph	Equation
A	$y = a(x - h)^2 + k$
B	$y = mx + c$
C	$y = p \times q^x + r$
D	$y = \dfrac{v}{x}$

d Graph **A** is a quadratic with vertex $(1, 5)$.

$\therefore \ h = 1$ and $k = 5$

When $x = 4, \ y = 0$

$\therefore \ a(4 - 1)^2 + 5 = 0$

$\therefore \ 9a = -5$

$\therefore \ a = -\frac{5}{9}$

Graph **B** is an exponential function with horizontal asymptote $y = -2$

$\therefore \ r = -2$

When $x = 0, \ y = 1$

$\therefore \ p \times q^0 - 2 = 1$

$\therefore \ p = 3$

When $x = 1, \ y = 4$

$\therefore \ 3 \times q^1 - 2 = 4$

$\therefore \ 3q = 6$

$\therefore \ q = 2$

Graph **C** is a linear function with gradient $\frac{2}{4} = \frac{1}{2}$ and y-intercept 2.

$\therefore \ m = \frac{1}{2}$ and $c = 2$

Graph **D** is a rectangular hyperbola passing through $(2, -2)$.

$\therefore \ -2 = \dfrac{v}{2}$

$\therefore \ v = -4$

SOLUTIONS TO TOPIC 7 (CALCULUS)

SHORT QUESTIONS

1 a $\qquad y = x^3 - 4.5x^2 - 6x + 13$

$\therefore \ \dfrac{dy}{dx} = 3x^2 - 2 \times 4.5x - 6$

$= 3x^2 - 9x - 6$

b When the gradient of the tangent is 6,

$\dfrac{dy}{dx} = 6$

$\therefore \ 3x^2 - 9x - 6 = 6$

$\therefore \ 3x^2 - 9x - 12 = 0$

$\therefore \ 3(x^2 - 3x - 4) = 0$

$\therefore \ 3(x - 4)(x + 1) = 0$

$\therefore \ x = 4$ or $x = -1$

So, the x-coordinates of the points are 4 and -1.

2 a $y = ax^2 + bx + c \quad \therefore \ \dfrac{dy}{dx} = 2ax + b$

b When $x = k, \ \dfrac{dy}{dx} = 0$

$\therefore \ 2ak + b = 0$

$\therefore \ 2ak = -b$

$\therefore \ k = -\dfrac{b}{2a}$

c **i** If $\dfrac{dy}{dx} < 0, \ y$ is decreasing.

ii When $x < k, \ \dfrac{dy}{dx} < 0$, and when $x > k, \ \dfrac{dy}{dx} > 0$.

$\therefore \ \dfrac{dy}{dx}$ has sign diagram:

$\therefore \ x = k$ is a local minimum.

3 a $y = \dfrac{7}{x^3} = 7x^{-3}$

b $\dfrac{dy}{dx} = 7(-3)x^{-4}$

$= \dfrac{-21}{x^4}$

4 a $y = \dfrac{2x^4 - 4x^2 - 3}{x}$

$= \dfrac{2x^4}{x} - \dfrac{4x^2}{x} - \dfrac{3}{x}$

$= 2x^3 - 4x - 3x^{-1}$

b $\dfrac{dy}{dx} = 2(3)x^2 - 4 - 3(-1)x^{-2}$

$= 6x^2 - 4 + \dfrac{3}{x^2}$

c At $x = -1, \ \dfrac{dy}{dx} = 6(-1)^2 - 4 + \dfrac{3}{(-1)^2}$

$= 6 - 4 + 3$

$= 5$

So, at $x = -1$, the gradient is 5.

5 **a** $f(x) = 3x^2 - 4x^{-1} + 7$

$$\therefore \quad f'(x) = 3(2)x - 4(-1)x^{-2}$$
$$= 6x + \frac{4}{x^2}$$

b $f'(2) = 6(2) + \dfrac{4}{2^2}$
$$= 13$$

6 **a** **i** $f(x)$ is increasing for $x \geqslant 0$.

ii $f(x)$ is decreasing for $x \leqslant 0$.

b $f'(x) = 0$ at $x = 0$ and $x = 2$.

c

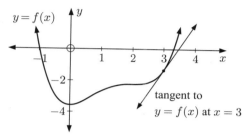

tangent to $y = f(x)$ at $x = 3$

7 **a** $f(x) = x^4 - 6x^2 - x + 3$

$\therefore \quad f'(x) = 4x^3 - 12x - 1$

b Maximum and minimum values of $f(x)$ occur when
$$f'(x) = 0$$
$$\therefore \quad 4x^3 - 12x - 1 = 0$$
$$\therefore \quad x \approx -1.69, \ -0.0835, \ 1.77$$
$$\{\text{using technology}\}$$

So, the x-coordinates of the points are -1.69, -0.0835, and 1.77.

8 **a** $g(x) = \left(3x - \dfrac{1}{x}\right)^2$
$$= (3x)^2 + 2(3x)\left(-\frac{1}{x}\right) + \left(\frac{1}{x}\right)^2$$
$$= 9x^2 - 6 + \frac{1}{x^2}$$

b $g(x) = 9x^2 - 6 + x^{-2}$
$$\therefore \quad g'(x) = 9(2)x + (-2)x^{-3}$$
$$= 18x - \frac{2}{x^3}$$

c The gradient of $g(x)$ at $x = -1$ is
$$g'(-1) = 18(-1) - \frac{2}{(-1)^3} = -18 + 2 = -16$$

So, the gradient is -16.

9 **a** $y = 2x^3 - 3x^2 - 264x + 13$
$$\therefore \quad \frac{dy}{dx} = 6x^2 - 6x - 264$$

b If $\dfrac{dy}{dx} = -12$ then
$$6x^2 - 6x - 264 = -12$$
$$\therefore \quad 6x^2 - 6x - 252 = 0$$
$$\therefore \quad 6(x^2 - x - 42) = 0$$
$$\therefore \quad 6(x - 7)(x + 6) = 0$$
$$\therefore \quad \text{the } x\text{-coordinates of the points are 7 and } -6.$$

10 If $y = x^2 + 1$ then $\dfrac{dy}{dx} = 2x$.

When $x = 2$, $y = 2^2 + 1 = 5$

and $\dfrac{dy}{dx} = 2(2) = 4$

$\therefore$ the gradient of the normal at $(2, 5)$ is $-\frac{1}{4}$.

$\therefore$ the equation of the normal is $\dfrac{y - 5}{x - 2} = -\frac{1}{4}$

which is $y - 5 = -\frac{1}{4}x + \frac{1}{2}$

Hence $y = -\frac{1}{4}x + \frac{11}{2}$

11 **a** $f(x) = 3x^3 + \dfrac{4}{x} - 3$
$$= 3x^3 + 4x^{-1} - 3$$
$$\therefore \quad f'(x) = 9x^2 - 4x^{-2}$$
$$= 9x^2 - \frac{4}{x^2}$$

b $f'(-2) = 9(-2)^2 - \dfrac{4}{(-2)^2}$
$$= 36 - 1$$
$$= 35$$

So, the gradient is 35.

c $f(-2) = 3(-2)^3 + \dfrac{4}{(-2)} - 3$
$$= -24 - 2 - 3$$
$$= -29$$

$\therefore$ the tangent has gradient 35 and passes through $(-2, -29)$.

It has the form $y = 35x + c$,

where $-29 = 35(-2) + c$

$\therefore \quad c = 41$

$\therefore$ the tangent is $y = 35x + 41$.

12 **a** $y = x^2 - 6x + 9$
$$\therefore \quad \frac{dy}{dx} = 2x - 6$$

b **i** At $x = a$, $\dfrac{dy}{dx} = -2$
$$\therefore \quad 2a - 6 = -2$$
$$\therefore \quad 2a = 4$$
$$\therefore \quad a = 2$$

ii At $x = 2$, $y = 2^2 - 6 \times 2 + 9 = 1$

So, the point is $(2, 1)$.

13 **a** $y = 3 - x + x^2 - 2x^3$
$$\therefore \quad \frac{dy}{dx} = -1 + 2x - 6x^2$$

When $x = -1$, $y = 3 - (-1) + 1 - 2(-1) = 7$

and $\dfrac{dy}{dx} = -1 - 2 - 6 = -9$

So, the tangent has gradient -9 and passes through $(-1, 7)$.

It has the form $y = -9x + c$

where $7 = 9 + c$

$\therefore \quad c = -2$

So, the tangent is $y = -9x - 2$

b Using technology, the tangent cuts the curve again at $(2.5, -24.5)$.

14 **a** $y = x^3 + ax + b$
$$\therefore \quad \frac{dy}{dx} = 3x^2 + a$$

b $\dfrac{dy}{dx} = 0$ when $x = 2$

$$\therefore \ 3(2)^2 + a = 0$$
$$\therefore \ a = -12$$

When $x = 2$, $y = -5$

$$\therefore \ 2^3 + (-12)(2) + b = -5$$
$$\therefore \ 8 - 24 + b = -5$$
$$\therefore \ b = 11$$

15 a
$$f(x) = \dfrac{x+1}{x^2}$$
$$= \dfrac{1}{x} + \dfrac{1}{x^2}$$
$$= x^{-1} + x^{-2}$$
$$\therefore \ f'(x) = -x^{-2} - 2x^{-3}$$
$$= -\dfrac{1}{x^2} - \dfrac{2}{x^3}$$

b We use technology to graph $y = f(x)$ on the domain $x < 0$. The local minimum is $(-2, -\frac{1}{4})$.

c

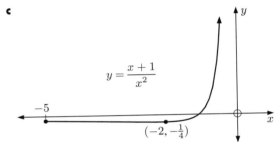

$$y = \dfrac{x+1}{x^2}$$

$(-2, -\frac{1}{4})$

16 a $(2x - 1)(x^2 - 1) = 2x^3 - 2x - x^2 + 1$

b $f(x) = 2x^3 - x^2 - 2x + 1$
$$\therefore \ f'(x) = 6x^2 - 2x - 2$$

c $f(-1) = 2(-1)^3 - (-1)^2 - 2(-1) + 1$
$$= -2 - 1 + 2 + 1$$
$$= 0$$

$f'(-1) = 6(-1)^2 - 2(-1) - 2$
$$= 6 + 2 - 2$$
$$= 6$$

$\therefore$ the normal at $(-1, 0)$ has gradient $= -\frac{1}{6}$

$\therefore$ the equation of the normal is $\dfrac{y - 0}{x + 1} = -\frac{1}{6}$

which is $y = -\frac{1}{6}x - \frac{1}{6}$.

17 a
$$f(x) = \dfrac{3}{x^2} + x - 4$$
$$= 3x^{-2} + x - 4$$
$$\therefore \ f'(x) = -6x^{-3} + 1$$
$$= -\dfrac{6}{x^3} + 1$$

b $f'(1) = -6 + 1 = -5$

This means that the tangent to the curve at the point where $x = 1$ has gradient -5.

c The gradient of the curve is zero when $f'(x) = 0$

$$\therefore \ -\dfrac{6}{x^3} + 1 = 0$$
$$\therefore \ \dfrac{6}{x^3} = 1$$
$$\therefore \ x^3 = 6$$
$$\therefore \ x = \sqrt[3]{6} \approx 1.82$$

18 a
$$f(x) = ax^2 + bx + c$$
$$\therefore \ f'(x) = 2ax + b$$

b $f'(x) = 2x - 6$

$\therefore$ by comparison with **a**, $\quad 2a = 2 \quad$ and $\quad b = -6$
$$\therefore \ a = 1 \quad \text{and} \quad b = -6$$

c The minimum value of f occurs when $f'(x) = 0$
$$\therefore \ 2x - 6 = 0$$
$$\therefore \ x = 3$$

d When $x = 3$, $f(x)$ has the minimum value 2.

$\therefore$ substituting into $f(x) = x^2 - 6x + c$, we have
$$(3)^2 - 6(3) + c = 2$$
$$\therefore \ 9 - 18 + c = 2$$
$$\therefore \ c = 11$$

19 a $v(t) = 7t - t^2 \ \text{km h}^{-1}$
$$\therefore \ v(1) = 7(1) - (1)^2 = 7 - 1 = 6 \ \text{km h}^{-1}$$

b $v(t) = 6 \ \text{km h}^{-1}$ when $7t - t^2 = 6$
$$\therefore \ 0 = t^2 - 7t + 6$$
$$\therefore \ 0 = (t - 6)(t - 1)$$

$\therefore \ v(t)$ is again $6 \ \text{km h}^{-1}$ when $t = 6$ hours.

c $v'(t) = 7 - 2t \ \text{km h}^{-2}$

d The maximum velocity occurs when $v'(t) = 0$
$$\therefore \ 7 - 2t = 0$$
$$\therefore \ t = \tfrac{7}{2}$$

Now $v\left(\tfrac{7}{2}\right) = 7\left(\tfrac{7}{2}\right) - \left(\tfrac{7}{2}\right)^2 = \tfrac{49}{2} - \tfrac{49}{4} = \tfrac{49}{4}$
$$= 12.25 \ \text{km h}^{-1}$$

$\therefore$ the maximum velocity is $12.25 \ \text{km h}^{-1}$.

20 a The gradient of the tangent through $A(6, 7)$ and $B(0, -5)$ is $m = \dfrac{y_2 - y_1}{x_2 - x_1} = \dfrac{7 - (-5)}{6 - 0} = 2$.

b
$$y = x^2 - kx + 11$$
$$\therefore \ \dfrac{dy}{dx} = 2x - k$$

c At $x = 4$, the gradient of the tangent is 2.
$$\therefore \ \dfrac{dy}{dx} = 2 \ \text{when} \ x = 4$$
$$\therefore \ 2(4) - k = 2 \quad \{\text{from } \textbf{b}\}$$
$$\therefore \ k = 6$$

21 a $f(x) = x^2 \quad \therefore \quad f(2) = 2^2 = 4$

b $f(2 + h) = (2 + h)^2$
$$= 4 + 4h + h^2$$

c $m = \dfrac{f(2 + h) - f(2)}{h}$
$$= \dfrac{4 + 4h + h^2 - 4}{h}$$
$$= \dfrac{h(4 + h)}{h}$$
$$= 4 + h$$

22 a $y = 4x^{-1} + x \quad \therefore \quad \dfrac{dy}{dx} = -4x^{-2} + 1$
$$= \dfrac{-4}{x^2} + 1$$

b $\dfrac{dy}{dx} = 0$ when $1 - \dfrac{4}{x^2} = 0$

$$\therefore \quad \frac{4}{x^2} = 1$$

$$\therefore \quad x^2 = 4$$

$$\therefore \quad x = \pm 2$$

c At $x = 2,\quad y = \dfrac{4}{2} + 2 = 4$

At $x = -2,\quad y = \dfrac{4}{-2} + (-2) = -4$

Sign diagram for $\dfrac{dy}{dx}$:

$\therefore$ there is a local minimum at $(2, 4)$
and a local maximum at $(-2, -4)$.

23 a $\quad H(t) = 80 - 5t^2$
$\therefore \quad H(0) = 80 - 5(0)^2 = 80$
The initial height is 80 m.

b The toy hits the ground when $H(t) = 0$

$$\therefore \quad 80 - 5t^2 = 0$$
$$\therefore \quad 80 = 5t^2$$
$$\therefore \quad t^2 = 16$$
$$\therefore \quad t = \pm 4$$
But $t > 0$, so $t = 4$ seconds.

So, the toy hits the ground after 4 seconds.

c $H'(t) = -10t$ m s^{-1}

d $H'(2) = -10(2) = -20$ m s^{-1}
After 2 seconds of flight, the toy aeroplane is travelling at 20 m s^{-1} towards the ground.

24 a $\dfrac{4 - 3x + x^2}{x^2} = \dfrac{4}{x^2} - \dfrac{3x}{x^2} + \dfrac{x^2}{x^2}$

$$= \frac{4}{x^2} - \frac{3}{x} + 1$$

b $\dfrac{d}{dx}\left(\dfrac{4}{x^2} - \dfrac{3}{x} + 1\right) = \dfrac{-8}{x^3} + \dfrac{3}{x^2}$

c $\dfrac{d}{dx}\left(\dfrac{4 - 3x + x^2}{x^2}\right) = 0$

when $\dfrac{-8}{x^3} + \dfrac{3}{x^2} = 0$

$$\therefore \quad \frac{-8 + 3x}{x^3} = 0$$

$$\therefore \quad 3x = 8$$

$$\therefore \quad x = \frac{8}{3}$$

25 a
$$y = ax^{-1} - x^2 + 1$$
$$\therefore \quad \frac{dy}{dx} = -ax^{-2} - 2x$$

But when $x = 2,\quad \dfrac{dy}{dx} = -5$

$$\therefore \quad \frac{-a}{(2)^2} - 2(2) = -5$$

$$\therefore \quad -\frac{a}{4} - 4 = -5$$

$$\therefore \quad -\frac{a}{4} = -1$$

$$\therefore \quad a = 4$$

b At $x = 2,\quad y = \dfrac{4}{2} - 2^2 + 1 = -1$
$\therefore$ the point is $(2, -1)$.

c The tangent at $x = 2$ has gradient $m = -5$
$\therefore$ its equation is $y = -5x + c$.
Substituting $(2, -1)$ gives $-1 = -10 + c$
$$\therefore \quad c = 9$$
$\therefore$ the tangent has equation $y = -5x + 9$.

26 a $s = 2t^3 - 18t^2 + 54t$
At $t = 0,\quad s = 0$ km
At $t = 4,\quad s = 2(4)^3 - 18(4)^2 + 54(4)$
$$= 56 \text{ km}$$
$\therefore$ the average velocity $= \dfrac{56}{4} = 14$ km h^{-1}

b $\dfrac{ds}{dt} = 6t^2 - 36t + 54$ km h^{-1}

c The hikers have stopped when $\dfrac{ds}{dt} = 0$

$$\therefore \quad 6t^2 - 36t + 54 = 0$$
$$\therefore \quad 6(t^2 - 6t + 9) = 0$$
$$\therefore \quad 6(t - 3)^2 = 0$$
$$\therefore \quad t = 3$$
Hence, the hikers stop momentarily after 3 hours.

d $s(3) = 2(3)^3 - 18(3)^2 + 54(3) = 54$
So, the hikers travel 54 km before stopping.

27 a $f(x) = x^3 - 3x^2 - 24x + 26$
$f'(x) = 3x^2 - 6x - 24$

b $\qquad$ If $f'(x) = 0$
then $3x^2 - 6x - 24 = 0$
$$\therefore \quad 3(x^2 - 2x - 8) = 0$$
$$\therefore \quad 3(x - 4)(x + 2) = 0$$
$$\therefore \quad x = 4 \text{ or } x = -2$$

c $f'(1) = 3(1)^2 - 6(1) - 24$
$$= -27$$
$\therefore$ the gradient of the normal to $y = f(x)$ at the point where $x = 1$, is $\dfrac{1}{27}$.

28 a $f(x) = \frac{2}{3}x^3 - 2x^2 - 5x + 10$
$\therefore \quad f'(x) = 2x^2 - 4x - 5$

b $f'(0) = -5$

c The tangent has gradient -5, so it has the form $y = -5x + c$.
Now $f(0) = 10$, so the tangent must pass through $(0, 10)$.
$$\therefore \quad c = 10$$
So, the tangent has equation $y = -5x + 10$.

d Using technology, the tangent intersects the curve again at $(3, -5)$.

29 a $f(x) = 2x^2 - 4x - 5$
$\therefore \quad f'(x) = 4x - 4$

b

x	-3	1	6
$f'(x)$	-16	0	20

$f'(-3) = 4(-3) - 4 = -16$
$f'(x) = 0$ when $4x - 4 = 0$
$$\therefore \quad x = 1$$

c **i** $f'(x) = 0$ when $x = 1$
Now $f(1) = 2 - 4 - 5 = -7$
$\therefore$ M is at $(1, -7)$.

ii The gradient to the left of M is negative, and to the right of M is positive.
$\therefore$ the point M$(1, -7)$ is a local minimum.

30
$$y = 3x^3 + 1$$
$$\therefore \quad \frac{dy}{dx} = 9x^2$$

If the gradient of the normal is $-\frac{1}{9}$ then $\frac{dy}{dx} = 9$
$$\therefore \quad 9x^2 = 9$$
$$\therefore \quad x^2 = 1$$
$$\therefore \quad x = \pm 1$$

When $x = -1$, $y = 3(-1)^3 + 1 = -2$
When $x = 1$, $y = 3(1)^3 + 1 = 4$

$\therefore$ the gradient of the normal is $-\frac{1}{9}$ at the points $(-1, -2)$ and $(1, 4)$.

31 **a**
$$f(x) = x + x^{-1}$$
$$\therefore \quad f'(x) = 1 - x^{-2}$$
$$= 1 - \frac{1}{x^2}$$

So, $f'(x) = 0$ when $1 - \frac{1}{x^2} = 0$
$$\therefore \quad x^2 = 1$$
$$\therefore \quad x = \pm 1$$
But $x > 0$, so $x = 1$

b The local minimum occurs at $x = 1$
$f(1) = 1 + \frac{1}{1} = 2$ so A is $(1, 2)$.

c The sum of a positive number and its reciprocal is at least 2.

d

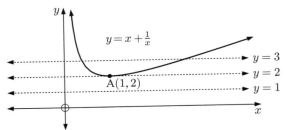

We consider the points of intersection of the function with the horizontal lines $y = 1$, $y = 2$, and $y = 3$.

i There are no points of intersection with $y = 1$, so there are no solutions.

ii There is one point of intersection with $y = 2$, so there is one solution.

iii There are two points of intersection with $y = 3$, so there are two solutions.

32 **a**

	A	B	C	D
$f(x)$	0	−	−	0
$f'(x)$	0	−	0	+

b

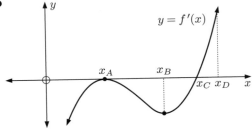

33 **a** $f(x)$ is defined provided $x + 3 \geqslant 0$
$\therefore$ the domain is $\{x \mid x \geqslant -3\}$

b $M(x) = \dfrac{y_B - y_A}{x_B - x_A} = \dfrac{f(x) - 8}{x - 1}$
$$= \frac{4\sqrt{x + 3} - 8}{x - 1}$$

c $p = M(2) = \dfrac{4\sqrt{5} - 8}{2 - 1} \approx 0.9443$ {to 4 d.p.}

$q = M(1.01) = \dfrac{4\sqrt{4.01} - 8}{1.01 - 1} \approx 0.9994$ {to 4 d.p.}

d From the table, we see that as point B gets closer and closer to point A, the gradient of the chord AB gets closer and closer to 1.
$\therefore$ the gradient $f(x)$ at point A is likely to be 1.

34
$$y = 2x^2 - \frac{3}{x} = 2x^2 - 3x^{-1}$$
$$\therefore \quad \frac{dy}{dx} = 4x + 3x^{-2}$$
$$= 4x + \frac{3}{x^2}$$

When $x = -0.5$, $y = 2(-\frac{1}{2})^2 - \dfrac{3}{(-\frac{1}{2})}$
$$= \tfrac{2}{4} + 6$$
$$= \tfrac{13}{2}$$

and $\dfrac{dy}{dx} = 4(-\frac{1}{2}) + \dfrac{3}{(-\frac{1}{2})^2}$
$$= -2 + 12$$
$$= 10$$

$\therefore$ the gradient of the normal at $(-\frac{1}{2}, \frac{13}{2})$ is $-\frac{1}{10}$

$\therefore$ the equation of the normal is
$$\frac{y - \frac{13}{2}}{x + \frac{1}{2}} = -\frac{1}{10}$$
which is $y - \frac{13}{2} = -\frac{1}{10}x - \frac{1}{20}$

Hence $y = -\frac{1}{10}x + \frac{129}{20}$

35 **a**
$$y = x^3 - 3x^2 - 24x + 7$$
$$\therefore \quad \frac{dy}{dx} = 3x^2 - 6x - 24$$

b If $\frac{dy}{dx} = 21$ then $3x^2 - 6x - 24 = 21$
$$\therefore \quad 3x^2 - 6x - 45 = 0$$
$$\therefore \quad 3(x^2 - 2x - 15) = 0$$
$$\therefore \quad 3(x - 5)(x + 3) = 0$$
$$\therefore \quad x = -3 \text{ or } 5$$

c Using **b**, we know $\frac{dy}{dx}$ is positive when $x = -3$ and 5.

We can therefore complete a sign diagram for $\frac{dy}{dx}$:

$$\xleftarrow{\qquad +\qquad}\underset{-2}{\quad}\xrightarrow{\;-\;}\underset{4}{\quad}\xrightarrow{\;+\;} x$$

So, $f(x)$ is decreasing for $-2 \leqslant x \leqslant 4$.

36 **a**
$$f(x) = 4x^2 + x^{-1}$$
$$\therefore \quad f'(x) = 8x - x^{-2}$$
$$= 8x - \frac{1}{x^2}$$

b $f'(x) = 0$ when $8x - \dfrac{1}{x^2} = 0$

$$\therefore \quad \frac{8x^3 - 1}{x^2} = 0$$

$$\therefore \quad 8x^3 = 1$$

$$\therefore \quad x^3 = \tfrac{1}{8}$$

$$\therefore \quad x = \tfrac{1}{2}$$

So, there can only be one maximum or minimum, and it occurs at $x = \tfrac{1}{2}$.

c $f(x)$ is decreasing for $x < 0$ and $0 < x \leqslant \tfrac{1}{2}$.

37 a $y = k^x$ is increasing for all $x \in \mathbb{R}$.

b y-intercept occurs when $x = 0$

$$\therefore \quad y = k^0$$

$$\therefore \quad y = 1 \quad \{k > 0\}$$

c The normal at $(0, 1)$ has gradient $m = -1$

$$\therefore \quad \text{its equation is} \quad y = -x + c$$

Substituting $(0, 1)$ gives $1 = c$

$$\therefore \quad \text{the normal is} \quad y = -x + 1$$

38 a
$$y = x^3 - x + 2$$

$$\therefore \quad \frac{dy}{dx} = 3x^2 - 1$$

At $x = -1$, $\dfrac{dy}{dx} = 3(-1)^2 - 1 = 2$

$\therefore$ the gradient of the tangent at $x = -1$ is 2.

b If the tangents are parallel, then when $x = k$, the gradient is 2.

Now $\dfrac{dy}{dx} = 2$

when $3x^2 - 1 = 2$

$$\therefore \quad x^2 = 1$$

$$\therefore \quad x = \pm 1$$

So, $k = 1$

39 a
$$f(x) = \frac{6}{x} - \frac{2}{x^3}$$

$$= 6x^{-1} - 2x^{-3}$$

$$\therefore \quad f'(x) = -6x^{-2} + 6x^{-4}$$

$$= \frac{-6}{x^2} + \frac{6}{x^4}$$

b $f'(x) = 0$ when $\dfrac{-6}{x^2} + \dfrac{6}{x^4} = 0$

$$\therefore \quad \frac{6}{x^4}(1 - x^2) = 0$$

$$\therefore \quad \frac{6}{x^4}(1 + x)(1 - x) = 0$$

$$\therefore \quad x = \pm 1$$

c At $x = 1$, $f(1) = 6 - 2 = 4$

$\therefore$ the tangent at $(1, 4)$ has gradient 0.

$\therefore$ the tangent is $y = 4$.

At $x = -1$, $f(-1) = -6 + 2 = -4$

$\therefore$ the tangent at $(-1, -4)$ has gradient 0.

$\therefore$ the tangent is $y = -4$.

40 a
$$s(t) = t^4 - 12t^3 + 48t^2 + 4t \text{ km}$$

$$\therefore \quad s'(t) = 4t^3 - 36t^2 + 96t + 4 \text{ km h}^{-1}$$

b $s'(3) = 4(3)^3 - 36(3)^2 + 96(3) + 4$

$$= 76 \text{ km h}^{-1}$$

After 3 hours, the truck is travelling at 76 km h^{-1}.

c $s(0) = 0$ km

$s(3) = (3)^4 - 12(3)^3 + 48(3)^2 + 4(3) = 201$ km

Hence $\dfrac{s(3) - s(0)}{3 - 0} = \dfrac{201 - 0}{3 - 0} = 67$ km h^{-1}.

d The result from **c** is an average speed over the first 3 hours.

41 a
$$f(x) = \tfrac{1}{3}x^3 - x^2 - 15x + 15\tfrac{2}{3}$$

$$\therefore \quad f'(x) = \tfrac{1}{3}(3x^2) - 2x - 15$$

$$= x^2 - 2x - 15$$

b
$$f'(x) = 0 \quad \text{when}$$

$$x^2 - 2x - 15 = 0$$

$$\therefore \quad (x - 5)(x + 3) = 0$$

$$\therefore \quad x = 5 \text{ or } x = -3$$

c
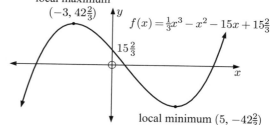

42 a $y = x^3 - 3x + 2$ $\therefore$ $\dfrac{dy}{dx} = 3x^2 - 3$

b Turning points occur when $\dfrac{dy}{dx} = 0$

$$\therefore \quad 3x^2 - 3 = 0$$

$$\therefore \quad 3(x^2 - 1) = 0$$

$$\therefore \quad 3(x + 1)(x - 1) = 0$$

$$\therefore \quad x = \pm 1$$

At $x = 1$, $y = (1)^3 - 3(1) + 2 = 0$

At $x = -1$, $y = (-1)^3 - 3(-1) + 2$

$$= -1 + 3 + 2$$

$$= 4$$

Sign diagram of $\dfrac{dy}{dx}$:

$\therefore$ there is a local maximum at $(-1, 4)$ and a local minimum at $(1, 0)$.

c
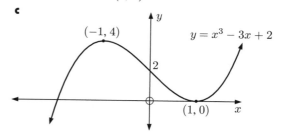

LONG QUESTIONS

1 a
$$f(x) = 4x^2 - 3x - \frac{2}{x}$$

$$\therefore \quad f(-1) = 4(-1)^2 - 3(-1) - \frac{2}{(-1)}$$

$$= 4 + 3 + 2$$

$$= 9$$

b
$$f(x) = 4x^2 - 3x - 2x^{-1}$$

$$\therefore \quad f'(x) = 4(2)x - 3 - 2(-1)x^{-2}$$

$$= 8x - 3 + \frac{2}{x^2}$$

c $f'(-1) = 8(-1) - 3 + \dfrac{2}{(-1)^2}$

$\qquad\qquad = -8 - 3 + 2$

$\qquad\qquad = -9$

So, the gradient of the tangent at $x = -1$ is -9.

d The tangent passes through $(-1, 9)$, and has gradient $m = -9$.

$\therefore$ equation is $y = -9x + c$.

Using the point $(-1, 9)$ gives $9 = -9(-1) + c$

$\qquad\qquad\qquad\qquad\qquad \therefore \quad 9 = 9 + c$

$\qquad\qquad\qquad\qquad\qquad \therefore \quad c = 0$

$\therefore$ equation of the tangent is $y = -9x$.

e Graphing $y = 4x^2 - 3x - \dfrac{2}{x}$ with $y = -9x$

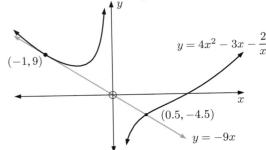

$y = 4x^2 - 3x - \dfrac{2}{x}$

$(-1, 9)$

$(0.5, -4.5)$

$y = -9x$

The tangent meets the curve again at $(0.5, -4.5)$.

2 a $C(x) = -0.2x^2 + 4x + 10$ for $0 \leqslant x \leqslant 10$

$C(5) = -0.2(5)^2 + 4(5) + 10 = \$25$

Hence, it costs $25 to produce 5 bracelets.

b $C'(x) = -0.2(2x) + 4$

$\qquad\quad = -0.4x + 4$

c If C is measured in dollars and x is a number of bracelets, then $\dfrac{dC}{dx}$ has units "dollars per bracelet".

d $C'(5) = -0.4(5) + 4$

$\qquad\quad = 2$ dollars per bracelet

e At $x = 5$, $\dfrac{dC}{dx} > 0$

$\therefore \quad C(x)$ is increasing at $x = 5$

f $C'(x) = 0$ when $-0.4x + 4 = 0$

$\qquad\qquad\qquad\qquad \therefore \quad -0.4x = -4$

$\qquad\qquad\qquad\qquad \therefore \quad x = 10$

$C'(x)$ has sign diagram:

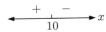

$\therefore$ there is a maximum when $x = 10$.

$C(10) = -0.2(10)^2 + 4(10) + 10 = 30$

$\therefore$ maximum value of $C(x)$ on $0 \leqslant x \leqslant 10$ is $30.

3 a Substituting $(1, 8)$ into $y = x^3 + ax^2 + bx + 3$

gives $(1)^3 + a(1)^2 + b(1) + 3 = 8$

$\qquad\qquad\qquad \therefore \quad a + b + 4 = 8$

$\qquad\qquad\qquad \therefore \quad a + b = 4$ .... (1)

b $\dfrac{dy}{dx} = 3x^2 + 2ax + b$

c i Tangent $y = 2x + 6$ has gradient $m = 2$.

ii At $x = 1$, $\dfrac{dy}{dx} = 2$

$\therefore$ using **b**, $3(1)^2 + 2a(1) + b = 2$

$\qquad\qquad\qquad\quad \therefore \quad 3 + 2a + b = 2$

$\qquad\qquad\qquad\qquad\quad \therefore \quad 2a + b = -1$ .... (2)

d We solve (1) and (2) simultaneously:

$\qquad\quad 2a + b = -1$

$\qquad\quad \underline{-a - b = -4} \qquad \{-1 \times (1)\}$

Adding: $\qquad a = -5$

Substituting $a = -5$ into (1) gives:

$\qquad\qquad -5 + b = 4$

$\qquad\qquad\quad \therefore \quad b = 9$

Hence, $a = -5$ and $b = 9$.

4 a $V(t) = 10t^2 - \frac{1}{3}t^3$ for $0 \leqslant t \leqslant 30$

$\therefore \quad V(5) = 10(5)^2 - \frac{1}{3}(5)^3$

$\therefore \quad V(5) \approx 208$ litres

After 5 minutes, the tank has 208 L of water in it.

b $V'(t) = 10(2)t - \frac{1}{3}(3)t^2$

$\qquad\quad = 20t - t^2$

$\qquad\quad = t(20 - t)$ L min^{-1}

c $V'(t) = t(20 - t)$

$\therefore \quad V'(t) = 0$ when $t = 0$ mins or $t = 20$ mins

d $V'(5) = 5(15) = 75$ L min^{-1}

$V'(25) = 25(-5) = -125$ L min^{-1}

e $V'(t)$ has sign diagram:

$+ \qquad - $

$0 \qquad 20 \qquad 30$ t (min)

$\therefore \quad V(t)$ is increasing for $0 \leqslant t \leqslant 20$.

5 a $f(x) = ax + \dfrac{b}{x}$

The tangent at $x = 1$ has equation $y = 3x + 2$.

Now, when $x = 1$, $y = 3(1) + 2 = 5$

$\therefore \quad (1, 5)$ is the point of contact.

b Since $f(1) = 5$,

$\therefore \quad a(1) + \dfrac{b}{(1)} = 5$

$\qquad \therefore \quad a + b = 5$ .... (1)

c $f'(x) = a - \dfrac{b}{x^2}$

d At $x = 1$, the gradient of the tangent $= 3$.

e Since $f'(1) = 3$, $a - b = 3$ .... (2)

f $a + b = 5$ .... (1)

$a - b = 3$ .... (2)

Solving simultaneously, $a = 4$, $b = 1$.

6 a Surface area $= 2\pi rh + 2\pi r^2$

$\therefore \quad 2\pi rh + 2\pi r^2 = 10\,000$

$\qquad \therefore \quad \pi rh = 5000 - \pi r^2$

$\qquad\qquad \therefore \quad h = \dfrac{5000 - \pi r^2}{\pi r}$

b Volume $V = \pi r^2 h$

$\qquad\qquad = \pi r^{\not2} \times \dfrac{5000 - \pi r^2}{\not\pi \not r}$

$\qquad\qquad = (5000 - \pi r^2)r$

$\qquad\qquad = 5000r - \pi r^3$ cm^3

c $\dfrac{dV}{dr} = 5000 - 3\pi r^2$

d $\dfrac{dV}{dr} = 0$ when $5000 - 3\pi r^2 = 0$

$$\therefore \ r^2 = \frac{5000}{3\pi}$$

$$\therefore \ r \approx 23.03 \quad \{\text{as } r > 0\}$$

$\dfrac{dV}{dr}$ has sign diagram:

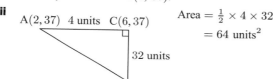

$\therefore$ the volume is maximised when the radius is about 23.03 cm.

e Substituting $r \approx 23.03$ into $V = 5000r - \pi r^3$ gives maximum volume $\approx 76\,800$ cm^3.

7 a $\qquad f(x) = x^3 - 12x^2 + 36x + 5$
$\therefore \ f'(x) = 3x^2 - 24x + 36$

b If $f'(x) = 0$ then $3x^2 - 24x + 36 = 0$
$$\therefore \ 3(x^2 - 8x + 12) = 0$$
$$\therefore \ 3(x-2)(x-6) = 0$$
$$\therefore \ x = 2 \text{ or } 6$$

c i $f(2) = 2^3 - 12(2)^2 + 36(2) + 5 = 37$
$\therefore$ A is $(2, 37)$.

ii $f(6) = 6^3 - 12(6)^2 + 36(6) + 5 = 5$
$\therefore$ B is $(6, 5)$.

d The tangent at A is horizontal, so P is $y = 37$.

e Since P is horizontal, Q is vertical.
$\therefore \ Q$ is $x = 6$.

f i P and Q intersect at C$(6, 37)$.

ii

A$(2, 37)$ 4 units C$(6, 37)$

32 units

B$(6, 5)$

Area $= \frac{1}{2} \times 4 \times 32$
$\qquad = 64$ units^2

iii $\tan(\widehat{CAB}) = \dfrac{\text{OPP}}{\text{ADJ}} = \dfrac{32}{4} = 8$

$\therefore \ \widehat{CAB} \approx 82.9°$

8 a Total perimeter $= 2x + y$
If 30 m of fence is available, then $2x + y = 30$
$$\therefore \ y = 30 - 2x$$

b Area $= xy$
$\therefore \ A(x) = x(30 - 2x)$

c If $A(x) = 100$, then $x(30 - 2x) = 100$
$$\therefore \ 30x - 2x^2 = 100$$
$$\therefore \ -2x^2 + 30x - 100 = 0$$
$$\therefore \ x^2 - 15x + 50 = 0$$
$$\therefore \ (x-10)(x-5) = 0$$
$$\therefore \ x = 10 \text{ or } x = 5$$

If $x = 10$, $\ y = 30 - 2(10) = 10$
If $x = 5$, $\quad y = 30 - 2(5) = 20$

Hence, $A(x) = 100$ m^2 when the dimensions are 10 m $\times$ 10 m or 5 m $\times$ 20 m.

d i $\qquad A(x) = 30x - 2x^2$
$\therefore \ A'(x) = 30 - 4x$
Now $A'(x) = 0$ when $30 - 4x = 0$
$$\therefore \ 4x = 30$$
$$\therefore \ x = 7.5$$

Sign diagram of $A'(x)$:

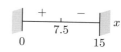

So, $x = 7.5$ m is a local maximum.
$\therefore \ y = 30 - 2(7.5) = 15$
So, the area is a maximum for a 7.5 m $\times$ 15 m enclosure.

ii $A = 7.5 \times 15 = 112.5$ m^2
So, the largest area possible is 112.5 m^2.

9 a $f(1) = (1)^3 = 1 \quad \therefore$ A has coordinates $(1, 1)$.

b B is located at $x = 1 + h$.

i When $h = 1$, $x = 2$
Now $f(2) = (2)^3 = 8$
$\therefore$ B is $(2, 8)$

ii When $h = 0.1$, $x = 1.1$
Now $f(1.1) = (1.1)^3 = 1.331$
$\therefore$ B is $(1.1, 1.331)$

iii When $h = 0.01$, $x = 1.01$
Now $f(1.01) = (1.01)^3 = 1.030\,301$
$\therefore$ B is $(1.01, 1.030\,301)$

c i When $h = 1$, $m_{AB} = \dfrac{8-1}{2-1} = 7$

ii When $h = 0.1$, $m_{AB} = \dfrac{1.331 - 1}{1.1 - 1} = 3.31$

iii When $h = 0.01$, $m_{AB} = \dfrac{1.030\,301 - 1}{1.01 - 1} = 3.0301$

d The smallest value of h, $h = 0.01$ would provide the best approximation to the tangent gradient at $x = 1$.

e $\qquad f'(x) = 3x^2$
$\therefore \ f'(1) = 3(1)^2 = 3$
$\therefore$ the actual tangent gradient is $m = 3$.

10 a Since the box has an open top,
the total surface area $= x^2 + 4xy$.

b Given that the surface area $= 1200$ cm^2,
$$\therefore \ x^2 + 4xy = 1200$$
$$\therefore \ 4xy = 1200 - x^2$$
$$\therefore \ y = \frac{1200 - x^2}{4x}$$

c $V = x^2 y = x^2 \left(\dfrac{1200 - x^2}{4x} \right)$

$\qquad = \dfrac{x(1200 - x^2)}{4}$

d From **c**, $V = 300x - \frac{1}{4}x^3$
$V'(x) = 300 - \frac{1}{4}(3x^2) = 300 - \frac{3}{4}x^2$

e Now, $V'(x) = 0$ when
$300 - \frac{3}{4}x^2 = 0$
$$\therefore \ \tfrac{3}{4}x^2 = 300$$
$$\therefore \ x^2 = \frac{4 \times 300}{3} = 400$$
$$\therefore \ x = 20 \quad \{\text{as } x > 0\}$$

Sign diagram for $V'(x)$:

$\therefore$ the maximum volume occurs when $x = 20$ cm.

When $x = 20$, $y = \dfrac{1200 - (20)^2}{4(20)} = 10$

So, the dimensions for the maximum volume box are 20 cm × 20 cm × 10 cm.

f The maximum volume $= 20 \times 20 \times 10$ cm^3

∴ the maximum volume is 4000 cm^3.

11 **a** Let the length of the shorter side be L m.

$$\text{Now} \quad y = 2L + x$$
$$\therefore \quad 2L = y - x$$
$$\therefore \quad L = \frac{y - x}{2}$$

b

$$\text{Area} = L \times x$$
$$\therefore \quad 450 = \frac{y - x}{2} \times x$$
$$\therefore \quad \frac{900}{x} = y - x$$
$$\therefore \quad y = x + \frac{900}{x}$$

c

$$y = x + 900x^{-1}$$
$$\therefore \quad \frac{dy}{dx} = 1 - 900x^{-2} = 1 - \frac{900}{x^2}$$

d $\dfrac{dy}{dx} = 0$ when $1 - \dfrac{900}{x^2} = 0$

$$\therefore \quad x^2 = 900$$
$$\therefore \quad x = 30 \quad \{as \ x > 0\}$$

Sign diagram for $\dfrac{dy}{dx}$:

So, y is minimised when $x = 30$.

e When $x = 30$, $y = 30 + \dfrac{900}{30}$

$$= 60$$

So, he has to buy 60 m of fencing.

f $L = \dfrac{60 - 30}{2} = 15$

So, the finished enclosure is 30 m × 15 m.

12 **a** $h(0) = 2.5$ m.   The ball was 2.5 m above the ground.

b $h(5) = 2.5 + 12.5 \times 5 - 2.6 \times 5^2$

$$= 0 \text{ m}$$

The ball has returned to ground level.

c $h(t) = 2.5 + 12.5t - 2.6t^2$ m

$$\therefore \quad \frac{dh}{dt} = 12.5 - 5.2t \text{ m s}^{-1}$$

d When $t = 0$, $\dfrac{dh}{dt} = 12.5$ m s^{-1}.

The student threw the ball at 12.5 m s^{-1} upwards.

e $\dfrac{dh}{dt} = 0$ when $12.5 - 5.2t = 0$

$$\therefore \quad t = \frac{12.5}{5.2} \approx 2.40$$

So, the ball reaches its maximum height after 2.40 seconds.

f $h(2.4) = 2.5 + 12.5 \times 2.4 - 2.6 \times 2.4^2$

$$\approx 17.5 \text{ m}$$

Since $h(2.4) < 18$ m, and this was the point of maximum height, the ball did not reach the height of the building.

This is a guide for **students only** and not necessarily official IB policy.

PAPER 1

Paper 1 consists of **15 short questions** each worth **6 marks**, for a total of **90 marks**.

Full marks are awarded for a correct answer, regardless of working. ('C' marks)

If an answer is incorrect, marks can be awarded for appropriate working **seen** ('M' marks) and for partially correct answers **seen** in the working ('A' marks).

Marks may also be awarded for reasoning. ('R' marks)

PAPER 2

Paper 2 consists of **long questions** varying from **10 to 25 marks** to total **90 marks**.

Working is **expected** to be shown in this paper. Correct answers on their own, may only be awarded the A marks, or just some of the A marks. The 'AG' notation is used in cases where the final answer is given and the student is expected to show the process by which the answer is obtained.

Graphic calculators are **expected to be used**. The allocation of marks in the mark scheme allows for this through 'G' marks. G marks apply in situations where a candidate has not shown their working.

In some questions, there is an expectation that a graphics calculator will be used exclusively. In these cases the marks scheme allocates G marks only. The syllabus document provides information regarding the topic areas where graphics calculator use is expected.

The general rule for students to follow is to **always** try to show their working where possible.

WORKING

The following presentations, seen on a student's script, would normally be awarded the marks for method ('M' marks) allocated in the mark scheme.

- **A substituted formula**

  Example:

  Find the amount of money in an account which pays nominal interest of 6.8% p.a. compounding monthly, if $3000 was placed in this account 4 years ago.

  Solution:

  $F_v = 3000 \left(1 + \frac{6.8}{1200}\right)^{12 \times 4}$ M1 A1 A1

  $= \$3934.74$ {using technology}   A1 (G3)

  (M1 for substituted formula,   A1 for correct monthly rate,   A1 for correct periods,
  A1 for correct answer.   G3 for correct answer with no working seen.)

- **A sketch taken from a graphics calculator screen**

  Example:

  Find the x-intercepts for the graph of $y = 3x^2 - 9x + 4$.

  Solution:

  We sketch the graph of the function, remembering to include axes labels and an indication of scale.

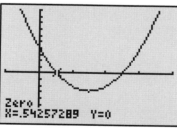

  The x-intercepts are $(0.543, 0)$ and $(2.46, 0)$ {using technology}   M1
  A1 A1 (G2)

- **A mathematical statement which indicates the candidate understands the nature of the solution**

  Example:

  Given $f'(x) = 6x - 9$, find the minimum value of the function $f(x) = 3x^2 - 9x + 4$.

  Solution:

  $$6x - 9 = 0$$ M1

  $$\therefore \ x = \frac{9}{6} \quad \left(\frac{3}{2} \text{ or } 1.5\right)$$ A1 (G1)

  (M1 for setting derivative to zero. An answer without working is awarded G1.)

  $$f(1.5) = 3(1.5)^2 - 9(1.5) + 4$$ M1

  Minimum value is $= -2.75$ A1ft (G2)

  (M1 is for substituting their value into $f(x)$. An answer without working is awarded full marks, G2.)

- **Follow through marks**

  Follow through marks (ft) are awarded in situations where an incorrect answer from the previous part is carried through to the subsequent part.

  If no additional errors are made, the student receives full marks for the subsequent part, provided working is shown.

  Possible follow through marks are indicated on the mark scheme.

- **Some Examination Reminders**

  ▶ Unless told otherwise in the question, all final answers should be given exactly, or else correct to 3 significant figures.

  ▶ For questions involving money, final answers can be given correct to 3 significant figures, to the nearest whole value, or to 2 decimal places.

  ▶ Your calculator display can be set for 3 significant figures or for various decimal places. It can also be set for scientific notation (standard form).

  ▶ Ensure your calculator is set to 'deg' for degrees. This will need to be done if your calculator is reset. For TI calculators, 'diagnostic' also needs to be set to 'on' if the calculator is reset.

  ▶ Ensure your calculator is set to 'trig' for graphs of periodic functions (sine and cosine).

  ▶ Use sensible view windows for sketches and general problem solving with graphs. Many students begin with "std" with a domain and range of -10 to $+10$. Negative values for the domain are often not appropriate for problems simulating real situations.

PAPER 1

1   **a** $5, -5, \sqrt{16}$ A2   **C2**

   b $\frac{1}{3}, 5, -5, \sqrt{16}, 0.\overline{6}$ A2   **C2**

   c $5, \sqrt{16}$ A2   **C2**

                                                                                                                              [6 marks]

2   **a**   **i** P(scores with both attempts) $= \frac{3}{5} \times \frac{3}{5}$ M1

                          $= \frac{9}{25}$ or 0.36   A1   **C2**

        ii P(misses at least once) $= 1 -$ P(scores with both attempts)

                          $= 1 - \frac{9}{25}$ M1

                          $= \frac{16}{25}$ or 0.64   A1ft   **C2**

   b Out of 30 shots, we expect the player to miss $30 \times \frac{2}{5}$ M1

                          $= 12$ times   A1   **C2**

                                                                                                                              [6 marks]

3   **a** Surface $= 9.85 \times 5.90$ M1

                   $= 58.115$ m^2

                   ≈ 58.1 m^2 A1   **C2**

   b Rounded measurements are 10 m by 6 m.

       Surface area $= 10 \times 6$

                   $= 60$ m^2 A1

       Percentage error $= \dfrac{60 - 58.115}{58.115} \times 100\%$ M1 A1ft

                   $\approx 3.24\%$ A1ft   **C4**

                                                                                                                              [6 marks]

4   **a**

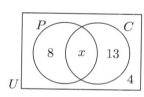

Total number of students $= 8 + 13 + 4 + x$

$\therefore \quad 30 = 25 + x$

$\therefore \quad x = 5$

$\therefore$ 5 students like both plain and chocolate milk.

                                                                                                                              A1   **C1**

   b

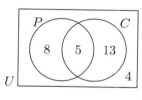

                                                                                                                              A1 A1 A1ft   **C3**

   c P(likes only one type of milk) $= \dfrac{8 + 13}{30}$

                   $= \frac{21}{30}$

                   $= \frac{7}{10}$ or 0.7   A1 A1ft   **C2**

                                                                                                                              [6 marks]

5   **a**   Total number of events $= 12 + 15 + 11 + 7 + 5$

                          $= 50$

       P(ticket costs more than \$60) $= \dfrac{11 + 7 + 5}{50}$

                          $= \frac{23}{50}$ or 0.46   A1   **C1**

b

Cost ($)	Number of events	Midpoint (x)
20 - 39	12	29.5
40 - 59	15	49.5
60 - 79	11	69.5
80 - 99	7	89.5
100 - 119	5	109.5
Total	50	

Using technology, estimates are: mean $\approx \$60.70$

standard deviation $\approx \$25.35$

A1
A1   **C2**

c 0.722 standard deviations above the mean is $\$60.70 + 0.722 \times \$25.35 = \$79.00$

A1ft

The percentage of events less than $79 = \dfrac{12 + 15 + 11}{50}$

$= \frac{38}{50} \times 100\%$

M1

$= 76\%$

A1ft   **C3**

[6 marks]

6 a 2000 USD $= 2000 \times 0.64$ GBP

M1

$= 1280$ GBP

A1   **C2**

b Amount remaining $= 1280 - 1100$

M1

$= 180$ GBP

A1ft   **C2**

180 GBP $= 180 \times \dfrac{1}{0.68}$ USD

M1

≈ 265 USD

A1ft   **C2**

[6 marks]

7 a $u_1 r^6 = 320$ and $u_1 r^9 = 2560$

M1

$\therefore \dfrac{u_1 r^9}{u_1 r^6} = \dfrac{2560}{320}$

$\therefore \quad r^3 = 8$

$\therefore \quad r = 2$

A1   **C2**

b $\quad u_1 2^6 = 320$

M1

$\therefore \quad 64 u_1 = 320$

$\therefore \quad u_1 = 5$

A1ft   **C2**

c $u_{20} = u_1 r^{19}$

M1

$= 5 \times 2^{19}$

$= 2\,621\,440$

A1ft   **C2**

[6 marks]

8 a The interest compounds monthly over 3 years.

$\therefore$ the total amount Ali needs to repay is $12\,000 \times \left(1 + \dfrac{8.5}{1200}\right)^{3 \times 12}$ euros

M1 A1

$\approx \text{€}15\,471.63$

A1ft   **C3**

b The car depreciates at 22.5% p.a. for 4 years.

$\therefore$ the value at the end of 2013 is $12\,000 \times \left(1 - \dfrac{22.5}{100}\right)^4$ euros

M1 A1

$\approx \text{€}4329.00$

A1   **C3**

[6 marks]

9 a $Q_3 = 77, \quad Q_1 = 65$

A1 A1   **C2**

b   i IQR $= Q_3 - Q_1$

$= 77 - 65$

$= 12$

A1ft

ii Range $=$ maximum $-$ minimum

$= 85 - 45$

A1

$= 40$

A1   **C3**

c 70 is the median value.

$\therefore$ 50% of the values are less than 70.

A1   **C1**

[6 marks]

10 **a** Profit = total sales − total cost

∴   for 100 boxes, profit = 12.50(100) − (9.5(100) + 45)   **M1**

= \$255   **A1**   **C2**

b The firm breaks even when $12.5x = 9.5x + 45$ **M1**

∴ $3x = 45$

∴ $x = 15$

So, 15 boxes must be produced and sold to break even.   **A1**   **C2**

c For the profit to be greater than \$1000, ∴ $12.50x − (9.5x + 45) > 1000$ **M1**

∴ $3x − 45 > 1000$

∴ $x > 348.3$

So, 349 boxes must be produced and sold.   **A1**   **C2**

[6 marks]

11 **a**

$p \wedge q$	$\neg(p \wedge q)$	$p \veebar q$	$\neg(p \wedge q) \vee q$	$(p \veebar q) \Rightarrow \neg(p \wedge q) \vee q$		
T	F	F	T	T	**A1**	
F	T	T	T	T	**A1**	
F	T	T	T	T	**A1**	
F	T	F	T	T	**A1**	**C4**

b If Bozo does not have a red nose then Bozo is not a clown.   **A1 A1**   **C2**

[6 marks]

12 A is $(-2, -3)$,   B is $(1, 3)$

a The gradient of AB $= \dfrac{3 - (-3)}{1 - (-2)} = 2$ **A1**

∴   the equation of AB is $y = 2x + c$

Substituting $(1, 3)$ gives $3 = 2(1) + c$

∴ $c = 1$ **M1**

The equation of AB is $y = 2x + 1$

or $2x - y + 1 = 0$ **A1ft**   **C3**

b Midpoint of AB is $\left(\dfrac{-2 + 1}{2}, \dfrac{-3 + 3}{2}\right)$, or $\left(-\tfrac{1}{2}, 0\right)$ **A1**

The gradient of the perpendicular bisector is $-\tfrac{1}{2}$ $\{$as $2 \times -\tfrac{1}{2} = -1\}$ **A1ft**

∴   its equation is $y = -\tfrac{1}{2}x + c$

Substituting $\left(-\tfrac{1}{2}, 0\right)$ gives $0 = -\tfrac{1}{2}\left(-\tfrac{1}{2}\right) + c$

∴ $c = -\tfrac{1}{4}$

∴   the equation of perpendicular bisector is $y = -\tfrac{1}{2}x - \tfrac{1}{4}$

or $4y = -2x - 1$

or $2x + 4y + 1 = 0$ **A1ft**   **C3**

[6 marks]

13 **a** $f(x) = ax^2 + bx + d$ ∴ $f'(x) = 2ax + b$ **A1**   **C1**

b $f'(x) = 5x - 10$

Equating coefficients gives $2a = 5$ and $b = -10$

∴ $a = 2.5$ and $b = -10$ **A1ft A1ft**   **C2**

c $f'(x) = 0$ when $x = 2$ **A1ft**

Now $f(2) = 2.5 \times 2^2 - 10 \times 2 + d$

$= d - 10$

∴ $d - 10 = -4$ **M1**

∴ $d = 6$ **A1ft**   **C3**

[6 marks]

14 **a**

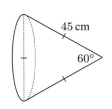

The triangle which bisects the cone is equilateral with sides 45 cm.

∴   the diameter of the megaphone is 45 cm.   **A1**   **C1**

Mathematical Studies SL – Exam Preparation & Practice Guide (3ʳᵈ edition)

b

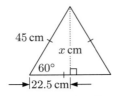

Let the height be x cm.

Now $\tan 60° = \dfrac{x}{22.5}$ M1

$\therefore \quad x \approx 39.0$ A1ft

$\therefore$ height of cone is 39 cm.

Volume of cone $= \frac{1}{3}\pi r^2 h$

$\approx \frac{1}{3}\pi \times 22.5^2 \times 39$ M1 A1ft

$\approx 20\,700 \text{ cm}^3$ A1ft   **C5**

[6 marks]

15   **a**   The top 33% of candidates scored a B or better.

If $P(X > k) = 0.33$

then $k \approx 57.3$ M1 A1

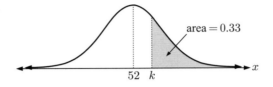

area $= 0.33$

$\therefore$ the minimum mark required for a B is 58.   A1   **C3**

b

$P(X < 40) \approx 0.159$ M1 A1

$\therefore$ the expected number of students scoring less than 40 is

300×0.159

$= 48$ students   A1   **C3**

[6 marks]

PAPER 2

1   **a**

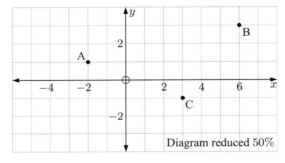

Diagram reduced 50%

   A1 A1   **2**

b   **i**   Gradient of BC $= \dfrac{-1 - 3}{3 - 6} = \dfrac{-4}{-3} = \frac{4}{3}$ M1 A1 (G2)

   ii   The opposite sides of a parallelogram are parallel, and parallel lines have the same gradient.

$\therefore$ gradient of AD $=$ gradient of BC   A1

   iii   Gradient of AD $= \dfrac{d - 1}{-5 - -2} = \dfrac{4}{3}$ M1

$\therefore \quad \dfrac{d - 1}{-3} = \dfrac{4}{3}$

$\therefore \quad d - 1 = \frac{4}{3} \times -3$

$\therefore \quad d - 1 = -4$

$\therefore \quad d = -3$ A1ft (G2)   **5**

c   **i**   Length AB $= \sqrt{(6 - -2)^2 + (3 - 1)^2}$ M1 A1

$= \sqrt{68}$ units   A1 (G2)

   ii $\cos A\widehat{B}C = \dfrac{\sqrt{68}^2 + 5^2 - \sqrt{29}^2}{2 \times \sqrt{68} \times 5}$ M1 A1ft

$\therefore \quad A\widehat{B}C \approx 39.1°$ A1ft (G2)   **6**

d   Area ABCD $= 2 \times$ area triangle ABC

$= 2 \times \frac{1}{2} \times 5 \times \sqrt{68}\sin A\widehat{B}C$ M1 A1ft

$= 26$ units^2 A1ft (G2)   **3**

[16 marks]

2 a Let B represent a basketball being chosen, F represent a football being chosen, and V represent a volleyball being chosen.

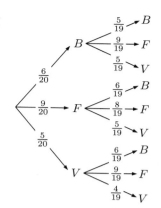

A4   **4**

b i P(two basketballs) = P(B then B)

$$= \tfrac{6}{20} \times \tfrac{5}{19}$$ M1

$$= \tfrac{30}{380} \quad (\approx 0.0789)$$ A1ft(G2)

ii P(a basketball and a football) = P(B then F) + P(F then B)

$$= \tfrac{6}{20} \times \tfrac{9}{19} + \tfrac{9}{20} \times \tfrac{6}{19}$$ M1 A1ft

$$= \tfrac{108}{380} \quad (\approx 0.284)$$ A1ft(G2)

iii P(both balls are the same) = P(B then B) + P(F then F) + P(V then V)

$$= \tfrac{6}{20} \times \tfrac{5}{19} + \tfrac{9}{20} \times \tfrac{8}{19} + \tfrac{5}{20} \times \tfrac{4}{19}$$ M1 A1ft

$$= \tfrac{122}{380} \quad (\approx 0.321)$$ A1ft(G2)   **8**

c With replacement:

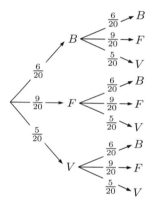

i P(two volleyballs) = P(V and V)

$$= \tfrac{5}{20} \times \tfrac{5}{20}$$ M1 A1ft

$$= \tfrac{1}{4} \times \tfrac{1}{4}$$

$$= \tfrac{1}{16}$$ A1ft(G2)   **3**

ii P(both B | the two balls are the same)

$$= \frac{\text{P}(B \text{ then } B)}{\text{P}(B \text{ then } B) + \text{P}(F \text{ then } F) + \text{P}(V \text{ then } V)}$$

$$= \frac{\tfrac{6}{20} \times \tfrac{6}{20}}{\tfrac{6}{20} \times \tfrac{6}{20} + \tfrac{9}{20} \times \tfrac{9}{20} + \tfrac{5}{20} \times \tfrac{5}{20}}$$ M1 A1ft

$$= \tfrac{18}{71}$$

$$\approx 0.254$$ A1ft(G2)   **3**

[18 marks]

3 a i $T_P(0) = 61 \times (0.95)^0 + 18$

$$= 61 + 18$$

$$= 79°\text{C} \qquad \therefore \ a = 79$$ A1

$$T_P(30) = 61 \times (0.95)^{30} + 18$$

$$\approx 31.1°\text{C} \qquad \therefore \ b \approx 31.1$$ A1   **2**

ii We need to solve $61 \times (0.95)^t + 18 = 25$ M1

So, $t \approx 42.2$ min   {using technology}   A1(G2)   **2**

b

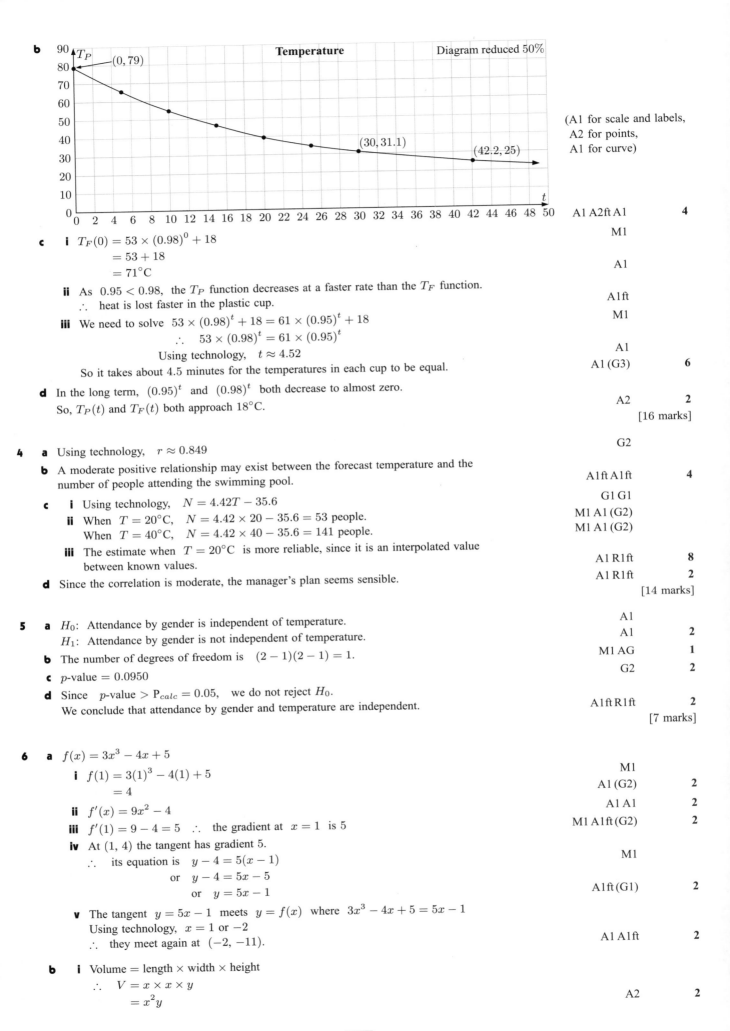

(A1 for scale and labels,
A2 for points,
A1 for curve)

A1 A2ft A1   **4**

c   **i** $T_F(0) = 53 \times (0.98)^0 + 18$ M1

$\qquad = 53 + 18$

$\qquad = 71°C$ A1

   ii   As $0.95 < 0.98$, the T_P function decreases at a faster rate than the T_F function.   A1ft

$\qquad \therefore$ heat is lost faster in the plastic cup.

  iii   We need to solve $53 \times (0.98)^t + 18 = 61 \times (0.95)^t + 18$ M1

$\qquad \therefore \quad 53 \times (0.98)^t = 61 \times (0.95)^t$ A1

$\qquad$ Using technology, $t \approx 4.52$ A1 (G3)   **6**

$\qquad$ So it takes about 4.5 minutes for the temperatures in each cup to be equal.

d   In the long term, $(0.95)^t$ and $(0.98)^t$ both decrease to almost zero.

So, $T_P(t)$ and $T_F(t)$ both approach $18°C$.   A2   **2**

[16 marks]

4   **a**   Using technology, $r \approx 0.849$ G2

  b   A moderate positive relationship may exist between the forecast temperature and the
number of people attending the swimming pool.   A1ft A1ft   **4**

  c   **i**   Using technology, $N = 4.42T - 35.6$ G1 G1

   ii   When $T = 20°C$, $N = 4.42 \times 20 - 35.6 = 53$ people.   M1 A1 (G2)

$\qquad$ When $T = 40°C$, $N = 4.42 \times 40 - 35.6 = 141$ people.   M1 A1 (G2)

  iii   The estimate when $T = 20°C$ is more reliable, since it is an interpolated value
between known values.   A1 R1ft   **8**

  d   Since the correlation is moderate, the manager's plan seems sensible.   A1 R1ft   **2**

[14 marks]

5   **a** H_0: Attendance by gender is independent of temperature.   A1

$\qquad H_1$: Attendance by gender is not independent of temperature.   A1   **2**

  b   The number of degrees of freedom is $(2-1)(2-1) = 1$.   M1 AG   **1**

  c p-value $= 0.0950$ G2   **2**

  d   Since p-value $>$ P$_{calc} = 0.05$, we do not reject H_0.

We conclude that attendance by gender and temperature are independent.   A1ft R1ft   **2**

[7 marks]

6   **a** $f(x) = 3x^3 - 4x + 5$

    i $f(1) = 3(1)^3 - 4(1) + 5$ M1

$\qquad = 4$ A1 (G2)   **2**

   ii $f'(x) = 9x^2 - 4$ A1 A1   **2**

  iii $f'(1) = 9 - 4 = 5$ $\therefore$ the gradient at $x = 1$ is 5   M1 A1ft (G2)   **2**

  iv   At $(1, 4)$ the tangent has gradient 5.

$\qquad \therefore$ its equation is $y - 4 = 5(x - 1)$ M1

$\qquad\qquad$ or $\quad y - 4 = 5x - 5$

$\qquad\qquad$ or $\quad y = 5x - 1$ A1ft (G1)   **2**

   v   The tangent $y = 5x - 1$ meets $y = f(x)$ where $3x^3 - 4x + 5 = 5x - 1$

$\qquad$ Using technology, $x = 1$ or -2

$\qquad \therefore$ they meet again at $(-2, -11)$.   A1 A1ft   **2**

  b   **i**   Volume = length $\times$ width $\times$ height

$\qquad \therefore \quad V = x \times x \times y$

$\qquad\qquad = x^2 y$ A2   **2**

ii $V = x^2 \left(\dfrac{30\,000 - x^2}{2x} \right)$ M1

$\quad = \tfrac{1}{2}x(30\,000 - x^2)$

$\quad = 15\,000x - \tfrac{1}{2}x^3$ A1ft(G2)   **2**

iii $\dfrac{dV}{dx} = 15\,000 - \tfrac{3}{2}x^2$ A1 A1ft   **2**

iv The maximum value occurs when $\dfrac{dV}{dx} = 0$ M1

$\qquad \therefore \quad \tfrac{3}{2}x^2 = 15\,000$

$\qquad \therefore \quad 3x^2 = 30\,000$

$\qquad \therefore \quad x^2 = 10\,000$

$\qquad \therefore \quad x = 100 \quad \{x > 0\}$ A1

$\therefore$ the maximum value of V is when $x = 100$.   A1ft(G2)   **3**

[19 marks]

SOLUTIONS TO TRIAL EXAMINATION 2

PAPER 1

1   **a**   **C** and **D**   A1 A1   **C2**

   b $0.0518 = 5.18 \times 10^{-2}$ A1 A1   **C2**

   c   Percentage error in rounding is

$\qquad \dfrac{0.0518 - 0.051\,762}{0.051\,762} \times 100\%$ M1

$\qquad \approx 0.0734\%$ A1   **C2**

[6 marks]

2   **a** $2 + 5 + 8 + 12 + 5 + 6 + 2 = 40$ sheep were weighed   A1   **C1**

   b   Mean weight

$\qquad = \dfrac{2 \times 10 + 5 \times 20 + 8 \times 30 + ... + 2 \times 70}{40}$ M1

$\qquad = \dfrac{1590}{40}$

$\qquad = 39.75$ kg   A1ft   **C2**

   c   150% of the mean weight is $39.75 \times 1.50 \approx 59.625$ kg

$\qquad$ Percentage of sheep going to market is $\dfrac{6+2}{40} \times 100\% = \tfrac{8}{40} \times 100\%$ M1

$\qquad\qquad\qquad\qquad\qquad\qquad\qquad\qquad\qquad\qquad = 20\%$ A1 A1ft   **C3**

[6 marks]

3   **a** $\cos 38° = \dfrac{DC}{10}$ M1

$\qquad\qquad\qquad\qquad 10 \times \cos 38° = DC$

$\qquad\qquad\qquad\qquad \therefore \quad DC \approx 7.88$ cm   A1   **C2**

   b $\sin 38° = \dfrac{AD}{10}$ M1

$\qquad 10 \times \sin 38° = AD$

$\qquad \therefore \quad AD \approx 6.16$ cm   A1

$\qquad$ Now $AB^2 = AD^2 + BD^2$

$\qquad \therefore \quad BD^2 = AB^2 - AD^2$

$\qquad \therefore \quad BD^2 = 8.5^2 - 6.16^2$ M1

$\qquad \therefore \quad BD = \sqrt{8.5^2 - 6.16^2}$

$\qquad \therefore \quad BD \approx 5.86$ cm   A1ft   **C4**

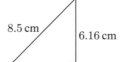

[6 marks]

4   **a**   The total revenue is $22 \times €42.50 = €935$.   A1   **C1**

   b   The cost of production is $C(22) = 15.6 \times 22 + 245$ M1

$\qquad\qquad\qquad\qquad\qquad\qquad = €588.20$ A1   **C2**

Mathematical Studies SL – Exam Preparation & Practice Guide (3rd edition)

c The profit is $935 - 588.20 = €346.80$

d The average profit per pair of jeans is $\dfrac{346.80}{22} = €15.76$

A1ft   **C1**

M1 A1ft   **C2**

[6 marks]

5   **a** $P = \{2, 3, 5, 7, 11, 13\}$

   b $Q = \{1, 2, 3, 4, 6, 8, 12\}$

   c $P \cup Q = \{1, 2, 3, 4, 5, 6, 7, 8, 11, 12, 13\}$

     $\therefore \;\; (P \cup Q)' = \{9, 10\}$

A1 A1   **C2**

A1 A1   **C2**

A1ft A1ft   **C2**

[6 marks]

6   In the tree diagram,
T = triangle, R = rectangle, H = rhombus.

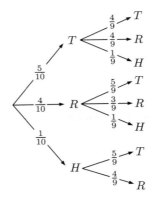

   a   P(both triangles) $= \dfrac{5}{10} \times \dfrac{4}{9} = \dfrac{2}{9}$

   b   P(one is a rhombus) $= \dfrac{5}{10} \times \dfrac{1}{9} + \dfrac{4}{10} \times \dfrac{1}{9} + \dfrac{1}{10} \times \dfrac{5}{9} + \dfrac{1}{10} \times \dfrac{4}{9}$

         $= \dfrac{1}{5}$

M1 A1   **C2**

M1 A1

A1ft   **C3**

   c   From the tree diagram, the probability that the second is a rectangle is $\dfrac{5}{10} \times \dfrac{4}{9} + \dfrac{4}{10} \times \dfrac{3}{9} + \dfrac{1}{10} \times \dfrac{4}{9} = \dfrac{36}{90}$

     The probability of selecting 2 rectangles is $\dfrac{4}{10} \times \dfrac{3}{9} = \dfrac{12}{90}$

     $\therefore$ the probability of the first being a rectangle is $\dfrac{12}{90} \div \dfrac{36}{90} = \dfrac{12}{36} = \dfrac{1}{3}$

A1   **C1**

[6 marks]

7   **a**

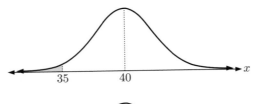

Let X be the weight in grams of a turtle egg.
$P(X < 35) \approx 0.073\,63$
The expected number of eggs which will weigh
less than 35 g is $0.073\,63 \times 90 \approx 6.6$ eggs.

M1 A1

A1   **C3**

   b

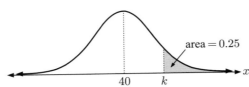

If $P(X < k) = 0.75$
    then $k \approx 42.327$
$\therefore$ the upper quartile ≈ 42.3 g

A1 M1

A1   **C3**

[6 marks]

8   **a**   The amount in the account after 3.5 years is

        $3000 \times \left(1 + \dfrac{7.5}{400}\right)^{3.5 \times 4}$

        $\approx \$3891.06$

M1 A1

A1   **C3**

   b   After 3.5 years, Jacinta withdraws half of the money, leaving $1945.53.

     The account balance will reach $4500 after n years, where $4500 = 1945.53 \times \left(1 + \dfrac{7.5}{400}\right)^{4n}$.

     Using technology, $n \approx 11.285$ years

     $\therefore$ the account balance will reach $4500 after 12 years.

M1 A1

A1   **C3**

[6 marks]

9   **a** $y = 2x^3 - 3x^2 - 264x + 13$

     $\therefore \;\; \dfrac{dy}{dx} = 2(3x^2) - 3(2x) - 264$

          $= 6x^2 - 6x - 264$

A1 A1 A1   **C3**

b The gradient is -12 if $\quad 6x^2 - 6x - 264 = -12$

$$\therefore \quad 6x^2 - 6x - 252 = 0$$
$$\therefore \quad 6(x^2 - x - 42) = 0$$
$$\therefore \quad 6(x - 7)(x + 6) = 0$$

So, the gradient is -12 when $x = -6$ and $x = 7$.

M1

A1ft A1ft   **C3**

[6 marks]

10   **a** $\quad 4r + 3p = 19 \quad \dots (1)$

A1   **C1**

    b $\quad 3r - 2p = 10 \quad \dots (2)$

A1   **C1**

    c Solving (1) and (2) simultaneously, $r = 4$, $p = 1$.

G3   **C3**

    d The cost of buying 5 reams of paper and 5 pens is $\quad 5 \times £4 + 5 \times £1 = £25.$

A1   **C1**

[6 marks]

11

	Climbing	Swimming	Mountain Bike	sum
Female	9	18	8	35
Male	15	16	24	55
sum	24	34	32	90

   a Expected values:

	Climbing	Swimming	Mountain Bike
Female	$\frac{24 \times 35}{90} \approx 9.33$	$\frac{34 \times 35}{90} \approx 13.2$	$\frac{32 \times 35}{90} \approx 12.4$
Male	$\frac{24 \times 55}{90} \approx 14.7$	$\frac{34 \times 55}{90} \approx 20.8$	$\frac{32 \times 55}{90} \approx 19.6$

M1 A1   **C2**

   b $\chi^2_{calc} \approx 5.44$

A1ft   **C1**

   c The p-value is 0.0658

A1ft   **C1**

   d Since $0.05 < 0.0658$ we do not have enough evidence to reject the hypothesis that the activities chosen are independent of gender.

R1 A1ft   **C2**

[6 marks]

12   **a** We are given $n = 20$, $S_n = 560$ and $u_n = 66$

$$S_n = \frac{n}{2}(u_1 + u_n)$$
$$\therefore \quad 560 = \frac{20}{2}(u_1 + 66)$$
$$\therefore \quad 560 = 10(u_1 + 66)$$
$$\therefore \quad 56 = u_1 + 66$$

M1 A1

So, the first term $u_1 = 56 - 66 = -10$

A1   **C3**

    b

$$\therefore \quad u_1 + (n-1)d = u_n$$
$$\therefore \quad u_1 + 19d = 66$$
$$\therefore \quad 19d = 76$$

M1 A1ft

The common difference is $d = 4$.

A1ft   **C3**

[6 marks]

13   **a** If x is a prime number then x is not a factor of 12.

A1   **C1**

    b If $x = 2$ or 3, p is true, but $\neg q$ is false.

So, $p \Rightarrow \neg q$ is false.

A1   **C1**

    c The inverse of $p \Rightarrow \neg q$ is $\neg p \Rightarrow q$.

In words: If x is not a prime number then x is a factor of 12.

A1 A1   **C2**

    d The inverse is not true for all values of x since 10 is not a prime number and it is not a factor of 12.

A1 A1   **C2**

[6 marks]

14   **a** At $x = 3$, $y = \dfrac{2^3}{3 - 2} = 8$

M1 A1   **C2**

    b When $x = 0$, $y = \dfrac{2^0}{0 - 2} = \dfrac{1}{-2} = -\frac{1}{2}$

M1

$\therefore$ y-intercept is $-\frac{1}{2}$.

A1   **C2**

    c $y \approx 7.54$ is the minimum value for $x \geqslant 2$

A1   **C1**

    d The vertical asymptote is $x = 2$, since this is when the denominator is zero.

A1   **C1**

[6 marks]

15 a

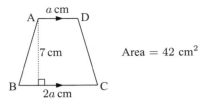

Area = 42 cm²

A1 A1   **C2**

b Area of trapezium $= \frac{1}{2}(a+b)h$

$\therefore \quad 42 = \frac{1}{2}(a+2a) \times 7$ M1 A1

$\therefore \quad 84 = 7 \times 3a$

$\therefore \quad 84 = 21a$

$\therefore \quad a = 4$ A1ft

The parallel sides of the trapezium have lengths 4 cm and 8 cm.   A1ft   **C4**

[6 marks]

PAPER 2

1 a i To increase an amount by 2.5%, we multiply it by 1.025.

We then add on the extra $500 on 1st January 2006.

$\therefore$ amount in bank $= 500(1.025) + 500$ M1 A1

$= 1012.50$ USD   A1 (G2)

ii The amount she will have one year later will be $(500(1.025) + 500) \times 1.025 + 500$ M1 A1 A1

$= 500(1.025)^2 + 500(1.025) + 500$

$= 1537.81$ USD   A1ft (G3)   7

b The amount after n years, the day before she makes her next deposit, will be

$500(1.025) + 500(1.025)^2 + ... + 500(1.025)^n$ M1

which is a geometric sequence with first term $500(1.025) = 512.50$ and common ratio 1.025

$\therefore$ the amount $= \dfrac{512.50 \times (1.025^n - 1)}{1.025 - 1}$ USD   A1ft (G2)   2

c At midnight on December 31st 2015, the amount will have been there for 11 years.

It is a day before Jaime puts another 500 USD in the bank.   A1

So, the amount will be $500(1.025) + 500(1.025)^2 + ... + 500(1.025)^{11}$ M1

$= 500(1.025) \times \left[\dfrac{(1.025)^{11} - 1}{1.025 - 1} \right]$

$= 6397.78$ USD   A1ft (G2)   3

d i Over 11 years Jaime invested $11 \times 500 = 5500$ USD   A1

ii Amount of interest earned is $6397.78 - 5500 = 897.78$ USD   A1ft   2

[14 marks]

2 a Using technology, the mean $= 1855 \div 30$ M1

≈ 61.8 A1ft (G2)   2

b i Median $= 62.5$ A1ft

ii Lower quartile $= 45$ A1ft

iii Upper quartile $= 75$ A1ft   3

c

```
  ┌───┬────┐
──┼───┤    ├──
  └───┴────┘
 0  10  20  30  40  50  60  70  80  90  100  %
```

(A1 for scale and label, A2 for box, A1 for whiskers)   A2ft A1ft A1ft   4

d i

Class interval	Frequency
30 - 39	2
40 - 49	7
50 - 59	4
60 - 69	6
70 - 79	7
80 - 89	4

A2ft   2

ii Modal groups are 40 - 49 and 70 - 79.   A1ft   1

e **i** Using technology, the mean for the grouped data ≈ 61.5

A2ft   **2**

    ii Error in estimating mean is $61.8 - 61.5 = 0.3$

$\therefore$ the percentage error $= \dfrac{0.3}{61.8} \times 100\% = 0.485\%$

M1 A1ft   **2**

[16 marks]

3 **a**

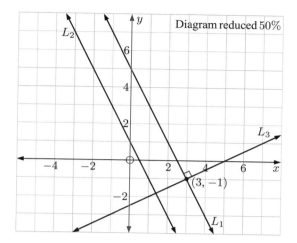

A4   **4**

b **i** $y = 5 - 2x = -2x + 5$ has a gradient of -2.

$y = kx + c$ has the same gradient $\therefore$ $k = -2$.

A1

Now $(2, -3)$ lies on $y = -2x + c$

$\therefore$ $-3 = -2(2) + c$

M1

$\therefore$ $c = 1$

So, $y = -2x + 1$

A1ft

    ii See the graph above.

A2   **5**

c **i** Substitution of $x = 3$ in L_1 gives $y = 5 - 2(3) = -1$ ✓

M1

So, $(3, -1)$ lies on the line.

A1 (G2)

    ii See the graph above.

A2   **4**

d **i** L_1 has gradient -2 which is $\frac{-2}{1}$

L_3 is perpendicular to L_1, so L_3 has gradient $\frac{1}{2}$

A1ft

Since $(3, -1)$ lies on L_3, the equation of L_3 is $\qquad y - -1 = \frac{1}{2}(x - 3)$

M1

$\therefore$ $2y + 2 = x - 3$

$\therefore$ $x - 2y - 5 = 0$

A1ft (G3)

    ii We have L_2: $2x + y = 1$ and L_3: $x - 2y = 5$

M1

Using technology to solve these simultaneously, we obtain $x = 1.4$ and $y = -1.8$.

So, the lines intersect at $(1.4, -1.8)$.

A1 A1ft (G3)   **6**

[19 marks]

4 **a** **i** When $x = 0$, $y = 2^0 = 1$

M1

So, B is at $(0, 1)$.

A1   **2**

    ii A and C are at the intersection of the curves.

So we need to solve $8 - x^2 = 2^{-x}$

M1

Using technology, the solutions are $x = -2$ and $x \approx 2.80$.

So, A is at $(-2, 4)$ and B is at $(2.80, 0.143)$.

A1 A1 (G3)   **3**

    iii When $2^{-x} > 8 - x^2$ the graph of $y = 2^{-x}$ is **above** the graph of $y = 8 - x^2$.

So, $x < -2$ or $x > 2.80$

A1 A1ft   **2**

b $\qquad y = 8 - x^2$

$\therefore \dfrac{dy}{dx} = -2x$

M1

When $x = -1$, the gradient of the tangent is $-2(-1) = 2$.

A1

$\therefore$ the equation of the tangent is $y = 2x + c$.

When $x = -1$, $y = 7$

$\therefore$ $2(-1) + c = 7$

$\therefore$ $c = 9$

$\therefore$ the equation of the tangent is $y = 2x + 9$.

M1 (AG)   **3**

c Using technology, the lines meet at $(-2.20, 4.60)$.

G1 G1   **2**

[12 marks]

5 **a**

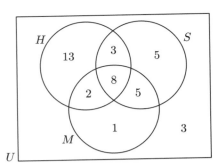

A1 A3ft   **4**

A1 A1ft

b   **i**   P(History only) $= \frac{13}{40}$

   ii   P(does not like Music) $= P(M') = \dfrac{13 + 3 + 5 + 3}{40}$

$= \frac{24}{40}$

A1ft

$= \frac{3}{5}$

   iii   P(M or S but not H) $= \dfrac{5 + 5 + 1}{40}$

$= \frac{11}{40}$

A1 A1ft

   iv   P(likes at least one) $= \frac{37}{40}$

A1ft   **6**

c   **i**   P($S \cap M$) $= \dfrac{8 + 5}{40}$

$= \frac{13}{40}$

A1 A1ft

   ii   P$\big((H \cap S \cap M)'\big) = \dfrac{40 - 8}{40}$

$= \frac{32}{40}$

$= \frac{4}{5}$

A1ft

   iii   P($H' \cap S'$) $= \dfrac{3 + 1}{40}$

$= \frac{4}{40}$

$= \frac{1}{10}$

A2ft   **5**

d   P($H \mid M$) $= \dfrac{2 + 8}{1 + 2 + 8 + 5}$

$= \frac{10}{16}$

$= \frac{5}{8}$

A1 A1ft   **2**

e   P(first likes H and S) $= \frac{11}{40}$

P(second likes H and S) $= \frac{10}{39}$

P(both like H and S) $= \frac{11}{40} \times \frac{10}{39}$

M1 A1

$= \frac{11}{156}$

A1ft (G3)   **3**

[20 marks]

6   **a**   The equation of the vertical asymptote is $x = 0$.

A1   **1**

   b $f(x) = 2x + 5 + \dfrac{3}{x}$

$= 2x + 5 + 3x^{-1}$

$\therefore \; f'(x) = 2 - 3x^{-2}$

$= 2 - \dfrac{3}{x^2}$

A1 A1 A1   **3**

   c   The gradient of the function at $x = \frac{1}{2}$ is

$f'(\tfrac{1}{2}) = 2 - \dfrac{3}{(\frac{1}{2})^2} = -10$

M1 A1 (G2)   **2**

d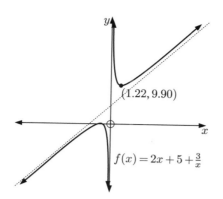

$(1.22, 9.90)$

$f(x) = 2x + 5 + \frac{3}{x}$

Using technology, the graph has a minimum at $(1.22, 9.90)$.

We see that $f'(x)$ is negative on $[\frac{1}{4}, \frac{3}{4}]$, so the function is decreasing on this interval.

G1

A1ft R1ft   **3**

[9 marks]

SOLUTIONS TO TRIAL EXAMINATION 3

PAPER 1

1 **a** $\dfrac{M}{EI} = \dfrac{6 \times 10^4}{(1.5 \times 10^8) \times (8 \times 10^2)}$ M1

$\qquad = 5 \times 10^{-7}$ A1 A1   **C3**

b $\sqrt[3]{\dfrac{34.5^2 - 103}{50.5 + 19}} \approx 2.501\,006\,789$

$\quad$ **i** 2.501   {to 4 significant figures}   A2

$\quad$ **ii** 3   {to nearest whole number}   A1ft   **C3**

[6 marks]

2 **a** Mean temperature $= \frac{202}{10}$ M1

$\qquad\qquad\qquad = 20.2°C$ A1   **C2**

b In order, the data is 15, 15, 18, 18, 19, 20, 22, 24, 25, 26.   M1
$\quad$ Median temperature $= 19.5°C$ A1   **C2**

c Percentage error is $\dfrac{20.2 - 20}{20.2} \times 100\%$ M1

$\qquad\qquad\qquad \approx 0.990\%$ A1   **C2**

[6 marks]

3 **a** $\qquad y = x^3 - 6x + 2$

$\quad \therefore \ \dfrac{dy}{dx} = 3x^2 - 6$ M1

$\quad$ When $x = 0$, $y = 2$ and $\dfrac{dy}{dx} = -6$.   A1

$\quad$ The gradient of the normal is $\frac{1}{6}$.

$\quad \therefore$ the equation of the normal is $y = \frac{1}{6}x + 2$.   M1 A1ft   **C4**

b The normal meets the x-axis when $\frac{1}{6}x + 2 = 0$ M1

$\qquad\qquad\qquad\qquad \therefore \ \frac{1}{6}x = -2$

$\qquad\qquad\qquad\qquad \therefore \ x = -12$

$\quad \therefore$ the point is $(-12, 0)$.   A1   **C2**

[6 marks]

4 **a** The line has gradient $= \dfrac{5 - 2}{1 - (-2)} = \dfrac{3}{3} = 1$ M1 A1

$\quad$ But $(1, 5)$ lies on the line, $\therefore$ it has equation $y - 5 = 1(x - 1)$ M1

$\qquad\qquad\qquad\qquad\qquad\qquad$ or $y = x - 1 + 5$

$\qquad\qquad\qquad\qquad\qquad\qquad$ or $y = x + 4$ A1ft   **C4**

b We need to solve simultaneously $y = 3 - x$ and $y = x + 4$.   M1
$\quad$ The solution is: $x = -0.5$ and $y = 3.5$.
$\quad$ So, the point of intersection is $(-0.5, 3.5)$.   A1ft   **C2**

[6 marks]

Mathematical Studies SL – Exam Preparation & Practice Guide (3ʳᵈ edition)

5 a

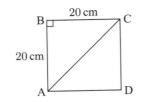
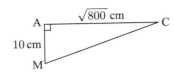

By Pythagoras' theorem
$$AC^2 = AB^2 + BC^2$$
$$\therefore \quad AC^2 = 20^2 + 20^2 = 800$$ M1
$$\therefore \quad AC = \sqrt{800}$$
$$\therefore \quad AC \approx 28.3 \text{ cm}$$ M1   **C2**

b

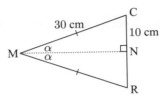

$$CM^2 = AC^2 + AM^2$$
$$\therefore \quad CM^2 = 800 + 10^2$$ M1
$$\therefore \quad CM^2 = 900$$
$$\therefore \quad CM = 30 \text{ cm}$$ A1ft   **C2**

c

Let N be the midpoint of CR
$$\sin\alpha = \frac{CN}{CM} = \frac{10}{30}$$ M1
$$\therefore \quad C\widehat{M}R = 2 \times \sin^{-1}\left(\tfrac{1}{3}\right)$$ A1ft   **C2**
$$\approx 38.9°$$
[6 marks]

6 $u_2 - u_1 = u_3 - u_2 = -7$ and so the series is arithmetic with common difference -7.

a Let n be the number of terms in the series.
$$\therefore \quad u_n = -48$$
But $u_n = u_1 + (n-1)d$
$$\therefore \quad -48 = 85 - 7(n-1)$$ M1 A1
$$\therefore \quad 7(n-1) = 133$$
$$\therefore \quad n - 1 = 19$$
$$\therefore \quad n = 20$$ A1   **C3**
So, the number of terms in the series is 20.

b Using $S_n = \frac{n}{2}(u_1 + u_n)$, $S_{20} = \frac{20}{2} \times (85 + -48)$ M1 A1
$$= 10 \times 37$$
$$= 370$$ A1ft   **C3**
[6 marks]

7 a The number of green beads is $\frac{1}{5}$ of $40 = 8$.   M1
There are 8 green beads in the bag and 32 yellow beads.   A1   **C2**

b $P(Y \text{ and } Y) = \frac{32}{40} \times \frac{31}{39}$ M1 A1
$$= \frac{124}{195}$$ A1ft   **C3**

c We need to find x such that $\dfrac{8+x}{40+x} = \dfrac{1}{2}$
$$\therefore \quad 16 + 2x = 40 + x$$
$$\therefore \quad x = 24$$
So, we need to add 24 green beads.   A1ft   **C1**
[6 marks]

8 a The factors of f are $(x+1)$ and $(x-3)$ M1
Now $(x+1)(x-3) = x^2 - 2x - 3$
and so $a = -2$, $b = -3$ A1 A1   **C3**

b $f(x) = x^2 - 2x - 3$, $\therefore$ $f'(x) = 2x - 2$ A1 A1ft   **C2**

c Minimum occurs when $f'(x) = 0$,
$$\therefore \quad 2x - 2 = 0$$
$$\therefore \quad x = 1.$$
When $x = 1$, $y = 1 - 2 - 3 = -4$
$$\therefore \text{ the minimum point is } (1, -4).$$ A1ft   **C1**
[6 marks]

9 a $x > 9$, so one possibility is $x = 10$.   A1   **C1**

b In order they are: $\underbrace{0\ 0\ 1\ 2\ 2\ 2}_{6}\ \underset{\underset{\text{median}}{\uparrow}}{3}\ \underbrace{4\ 4\ 4\ 4\ 4\ 4}_{6}$

$$\therefore\ y = 6$$

A2   **C2**

c The mean mark is 1, so $\dfrac{\sum fx}{\sum f} = 1$

$$\therefore\ \frac{0+4+6+6+4}{z+4+3+2+1} = 1$$

M1 A1

$$\therefore\ \frac{20}{z+10} = 1$$

$$\therefore\ z = 10$$

A1   **C3**

[6 marks]

10 **a** $X\widehat{Z}Y = (180 - 80 - 65)° = 35°$

Using the sine rule, $\dfrac{XZ}{\sin 65°} = \dfrac{5}{\sin 35°}$

M1 A1

$$\therefore\ XZ = \frac{5 \times \sin 65°}{\sin 35°}$$

$$XZ \approx 7.90 \text{ km}$$

A1   **C3**

b Using the cosine rule, $XW^2 = XY^2 + WY^2 - 2(XY)(WY)\cos 50°$

$$= 5^2 + 6^2 - 2 \times 5 \times 6 \times \cos 50°$$

M1 A1

$$XW \approx 4.74 \text{ km}$$

A1   **C3**

[6 marks]

11 **a** The axis of symmetry is midway between the x-intercepts.

$$\therefore\ \text{its equation is}\ \ x = \frac{-1+2}{2}$$

A1

which is $x = \frac{1}{2}$

A1   **C2**

b Since the x-intercepts are -1 and 2, the function has the form
$g(x) = a(x+1)(x-2)$ for some constant a.

M1

Since the y-intercept is 5, $a(1)(-2) = 5$

M1

$$\therefore\ a = -\tfrac{5}{2}$$

A1

Hence $g(x) = -\tfrac{5}{2}(x+1)(x-2)$

A1   **C4**

[6 marks]

12 **a** Using technology, $y \approx 1.12x - 12.9$

G1 G1   **C2**

b When $x = 60$, $y \approx 1.12 \times 60 - 12.9$

M1

$$\therefore\ y \approx 54.0$$

A1 (G2)   **C2**

c The estimate is not reliable since it is an extrapolation outside the known data range.

A1 R1   **C2**

[6 marks]

13 **a**

p	q	$\neg q$	$p \vee \neg q$
T	T	F	T
T	F	T	T
F	T	F	F
F	F	T	T

A2   **C2**

b

p	q	$\neg q$	$p \wedge q$	$(p \wedge q) \vee \neg q$
T	T	F	T	T
T	F	T	F	T
F	T	F	F	F
F	F	T	F	T

A2   **C2**

c $p \vee \neg q$ and $(p \wedge q) \vee \neg q$ are logically equivalent since their truth table columns are the same.

A1 A1   **C2**

[6 marks]

14 a

$$f(x) = \frac{10}{x} - x^2 = 10x^{-1} - x^2$$

$$\therefore \; f'(x) = 10(-1)x^{-2} - 2x$$

          A1

$$= -10x^{-2} - 2x$$

$$= \frac{-10}{x^2} - 2x$$

          A1   **C2**

b

$$f(-2) = \frac{10}{-2} - (-2)^2 = -9$$

          A1

and $f'(-2) = \dfrac{-10}{(-2)^2} - 2(-2) = \frac{3}{2}$

          M1

$\therefore$ the gradient of the normal at $(-2, -9)$ is $-\frac{2}{3}$.

          A1

$\therefore$ the equation of the normal is $\dfrac{y+9}{x+2} = -\frac{2}{3}$

which is $y + 9 = -\frac{2}{3}x - \frac{4}{3}$

or $y = -\frac{2}{3}x - \frac{31}{3}$

          A1   **C4**

[6 marks]

15 a Volume $= \pi r^2 h$

$$= \pi \times 4^2 \times 50 \text{ cm}^3\text{s}^{-1}$$

$$\approx \frac{16\pi \times 50 \times 3600}{1000} \text{ L h}^{-1}$$

          M1

$$\approx 9050 \text{ L h}^{-1}$$

          A1   **C3**

b Volume of swimming pool $= 10 \times 4.50 \times 1.50 \text{ m}^3$

$$= 67.5 \text{ m}^3$$

$$\equiv 67.5 \times 1000 \text{ L}$$

$$= 67\,500 \text{ L}$$

          M1

$\therefore$ the time to fill the pool is $\dfrac{67\,500}{9050} \approx 7.46$ hours

          M1 A1   **C3**

[6 marks]

PAPER 2

1 a i

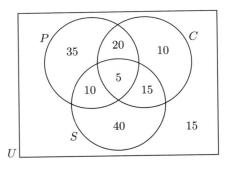

          A4   **4**

ii $150 - 135 = 15$ students are not members of any of the clubs.

          A1ft   **1**

iii $P(P \text{ only}) = \frac{35}{150}$

$$= \frac{7}{30} \quad (\approx 0.233)$$

          A1ft A1   **2**

iv $P(P \text{ or } C, \text{ but not both}) = \dfrac{35 + 10 + 10 + 15}{150}$

          M1

$$= \frac{70}{150}$$

$$= \frac{7}{15} \quad (\approx 0.467)$$

          A1ft (G1)   **2**

v $P(S \mid C) = \dfrac{5 + 15}{20 + 5 + 15 + 10}$

$$= \frac{20}{50}$$

$$= \frac{2}{5} \quad (= 0.4)$$

          A1 A1ft   **2**

b i H_0: Belonging to a club is independent of gender.

          A1   **1**

ii

	Performing Arts	Choir	Sports	sum
Males	40	30	80	150
Females	20	50	30	100
sum	60	80	110	250

The expected value for females belonging to the sports club is $\dfrac{100 \times 110}{250} = 44$ — A1 A1 AG   **2**

iii $\chi^2_{calc} \approx 25.4$ with p-value $\approx 3.04 \times 10^{-6}$ — G2   **2**

iv $\chi^2_{crit} < \chi^2_{calc}$ and the p-value < 0.05.

So, we reject H_0, and conclude that belonging to a club is not independent of gender. — A1ft R1ft   **2**

[18 marks]

2 a Number affected after 1 day is $10 \times 1.2 = 12$. — M1 A1 (G2)   **2**

b Number affected after 1 week is $10 \times (1.2)^7 \approx 35.8$ — M1 A1ft (G2)

≈ 36 — A1ft (G1)   **3**

c **i** $N = 10$ — A1ft

ii $a = 1.2$ — A2ft   **3**

d

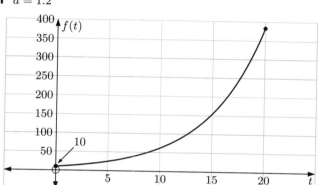

— A4ft   **4**

e $f(0) = 10 \times 1.2^0 = 10$

$f(20) = 10 \times (1.2)^{20} \approx 383$ {nearest integer}

The range is the interval $10 \leqslant y \leqslant 383$. — A1 A1 A1ft   **3**

f The horizontal asymptote is $y = 0$. — A1 A1ft   **2**

[17 marks]

3 a Median temperature for Nice $= 24°C$ — A1   **1**

b The lower quartile for Nice $= 23.5°C$ — A1   **1**

c Range for Nice $= 26 - 21 = 5°C$ — A1

Interquartile range $= 25 - 23.5 = 1.5°C$ — A2   **3**

d On the whole, Nice is hotter as the median temperature is higher than that of Geneva. — A1   **1**

e The middle half of the days have temperatures more spread out in Geneva. — A1 A1   **2**

f **i** There are 30 days in September. — A1

There are 25% of $30 = 7$ days on or above the upper quartile. — A1

Nice maximum = Geneva upper quartile value.

Smallest number of days $= 7$ — A1

ii Difference in maximum temperature on hottest day $= 28 - 26 = 2°C$. — A2   **5**

[13 marks]

4 a **i** Amount after 5 years $= 10\,000(1.035)^5$ — M1 A1

$\approx 11\,876.86$ — A1

$\approx 11\,900$ AUD — AG   **3**

ii Niki's money is doubled when $10\,000(1.035)^n = 20\,000$ — M1 A1

Using technology, $n \approx 20.149$

So, 21 years are needed. — A1 (G2)   **3**

b If the bank pays $r\%$ per annum, we need to solve: $\left(1 + \dfrac{r}{100}\right)^{12} = 2$ — M1

Using technology, $r \approx 5.9463$

$\therefore \ r \approx 5.95\%$ — (G3)   **4**

c After 6 years Sami will have $5000\left(1 + \frac{3}{100 \times 4}\right)^{6 \times 4}$

$$\approx 5982.07 \text{ AUD}$$

Interest $\approx 5982.07 - 5000$

$$\approx 982 \text{ AUD}$$

M1 A1
M1
M1 (G3)   **4**

d The cost of 5000 EUR is $= 5000 \times 1.25$

$$= 6250 \text{ AUD}$$

A commission of 1.5% is charged, so they need $6250 \div 0.985 = 6345.18$ AUD.

M1 A1ft (G2)   **4**
[18 marks]

---

5 a $f(0) = 4 - 2.5^0 = 3$

∴ the y-intercept is 3.

A1   **1**

b $f(x) = 0$ when $4 - 2.5^x = 0$

∴ $x \approx 1.51$ {using technology}

M1 A1   **2**

c

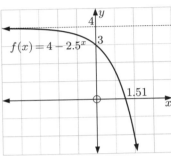

Diagram reduced 50%

A1 A1 A1 A1   **4**

d The horizontal asymptote is $y = 4$.

A1 A1   **2**

e Using technology, the graph meets the line $y = 3x - 2$ at the point $(1.09, 1.28)$.

A1 A1   **2**
[11 marks]

---

6 a PQ $= (20 - 2x)$ cm where $0 < x < 10$

A2   **2**

b Area of square base is $(20 - 2x)^2 = 400 - 80x + 4x^2$ cm^2

Volume of box with height x cm is $V = x(400 - 80x + 4x^2)$

$$= (4x^3 - 80x^2 + 400x) \text{ cm}^3$$

M1 A2
AG   **3**

c $\frac{dV}{dx} = 12x^2 - 160x + 400$

A1 A1 A1   **3**

d Using technology, $\frac{dV}{dx} = 0$ if $x = \frac{10}{3}$ or 10

M1 A1 A1 (G3)   **3**

e $x = 10$ is not possible.

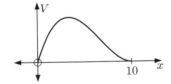

The maximum volume occurs if $x = 3\frac{1}{3}$.

A1ft A1ft   **2**
[13 marks]

SOLUTIONS TO TRIAL EXAMINATION 4

PAPER 1

1 $\dfrac{\sqrt{2.068}}{1.203 \times 0.0237} \approx 50.438\,399\,11$

a 50.44 {to 2 decimal places}

A1   **C1**

b 50 {to the nearest integer}

A1ft   **C1**

c 50.4384 {to 6 significant figures}

A1ft   **C1**

d 50.4384 {to nearest ten thousandth}

A1ft   **C1**

e $\approx 5.04 \times 10^1$

A1ft A1ft   **C2**
[6 marks]

2 a i If Joshua did not study hard at Mathematics then Joshua did not pass Mathematics.   A1 A1

ii Joshua did not study hard at Mathematics and Joshua passed Mathematics.   A1ft A1   **C4**

b $q \Rightarrow p$ A1

c the converse   A1ft   **C2**

[6 marks]

3 a Euro : Baht $= 1 : 37.794$

Maria receives $2400 \times 37.794 \times 0.985$ M1 A1

$\approx 89\,345$ Baht   A1   **C3**

b Maria has 10% of the Baht left, which is 8934.5 Baht.   A1ft

Since the exchange rate is now $1 : 36.481,$ she gets

$8934.5 \times \dfrac{1}{36.481} \times 0.985 \approx 241.23$ Euro   M1 A1ft   **C3**

[6 marks]

4 a

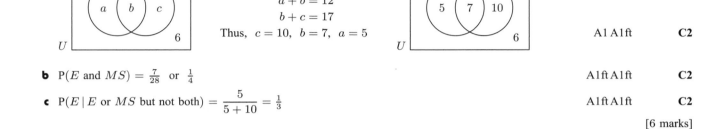

$a + b + c + 6 = 28$

$\therefore \quad a + b + c = 22$

$a + b = 12$

$b + c = 17$

Thus, $c = 10, \quad b = 7, \quad a = 5$

A1 A1ft   **C2**

b $P(E \text{ and } MS) = \frac{7}{28}$ or $\frac{1}{4}$ A1ft A1ft   **C2**

c $P(E \mid E \text{ or } MS \text{ but not both}) = \dfrac{5}{5 + 10} = \frac{1}{3}$ A1ft A1ft   **C2**

[6 marks]

5

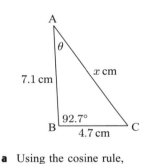

a Using the cosine rule,

$x^2 = 7.1^2 + 4.7^2 - 2 \times 7.1 \times 4.7 \times \cos 92.7°$ M1 A1

$\therefore \quad x^2 \approx 75.64$

$\therefore \quad x \approx 8.697$

So, AC ≈ 8.70 cm long.   A1   **C3**

b Using the sine rule,

$\dfrac{\sin \theta}{4.7} \approx \dfrac{\sin 92.7°}{8.697}$ M1 A1ft

$\therefore \quad \sin \theta \approx \dfrac{4.7 \times \sin 92.7°}{8.697} \approx 0.5398$

$\therefore \quad \theta \approx 32.7°$

So, $\widehat{CAB}$ is about $32.7°$.   A1ft   **C3**

[6 marks]

6 $P = 360 \times (1.045)^t$

a When $t = 0, \quad P = 360 \times (1.045)^0$

$= 360$ people

The population in 1960 was about 360 people.   A1   **C1**

b In 2005, $t = 45$

$\therefore \quad P = 360 \times (1.045)^{45} \approx 2610$ people   M1

The population in 2005 was about 2610 people.   A1   **C2**

c When $P = 800, \quad 800 = 360 \times (1.045)^t$ M1

Using technology, $t \approx 18.1$ A1

The population reached 800 in 1979.   A1ft   **C3**

[6 marks]

Mathematical Studies SL – Exam Preparation & Practice Guide (3rd edition)

7 **a**

Number of letters	1	2	3	4	5	6	7
Number of words	3	11	12	11	5	5	4

A2   **C2**

b

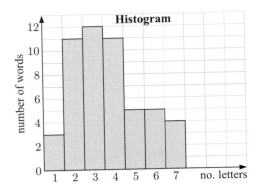

A4   **C4**
[6 marks]

8 **a** $I = 2500 \left(1 + \frac{3.2}{100}\right)^5 - 2500$

≈ 426 JPY

M1 A1
A1   **C3**

b $I = C \times \left(1 + \frac{r}{100k}\right)^{kn} - C$

$= 2500 \times \left(1 + \frac{3.2}{400}\right)^{4 \times 5} - 2500$

≈ 432 JPY

M1 A1
A1   **C3**
[6 marks]

9 The y-intercept is 6, so $c = 6$.

$f(1) = -1.5$

$\therefore \quad a + b + 6 = -1.5$

$\therefore \quad a + b = -7.5$ .... (1)

$f(-2) = 36$

$\therefore \quad a(4) + b(-2) + 6 = 36$

$\therefore \quad 4a - 2b = 30$

$\therefore \quad 2a - b = 15$ .... (2)

Solving (1) and (2) simultaneously, $a = 2.5$ and $b = -10$.

A1

M1 A1

M1
A1ft A1ft   **C6**
[6 marks]

10 **a** **i** 5 of the 58 are long and light.

$\therefore \quad$ P(long and light) $= \frac{5}{58}$

ii There are 41 lights of which $21 + 5$ are not medium.

$\therefore \quad$ P(not medium | light) $= \frac{26}{41}$

b P(both short and heavy) $= \frac{2}{58} \times \frac{1}{57} = \frac{1}{1653}$

A1 A1   **C2**

A1 A1ft   **C2**

M1 A1   **C2**
[6 marks]

11 **a** The profit forms a geometric sequence with $u_1 = 50\,000$, $r = 1.015$, and $u_{10} = u_1 \times r^9$

$= 50\,000 \times (1.015)^9$

$\approx 57\,200$ MAR

M1
A1   **C2**

b Total profit $= u_1 + u_2 + u_3 + ... + u_{10}$

$= \frac{50\,000(1.015^{10} - 1)}{1.015 - 1}$

$\approx 535\,136.08$

$\approx 535\,000$ MAR

M1

M1

A1
A1ft   **C4**
[6 marks]

12

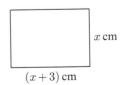

x cm

$(x + 3)$ cm

a Length $= (x + 3)$ cm

A1   **C1**

b Area = length × width

$\therefore \quad x(x+3) = 108$ A1ft   **C1**

c $\qquad x^2 + 3x - 108 = 0$

$\therefore \quad (x+12)(x-9) = 0$

$\qquad\qquad \therefore \quad x = -12 \text{ or } 9$ A1ft A1ft

$x > 0$ as lengths are positive

$\therefore \quad x = 9$ A1ft   **C3**

d So, the rectangle is 12 cm by 9 cm.   A1ft   **C1**

[6 marks]

13 a Value = 80% of $45 000

$= 0.8 \times \$45\,000$ M1

$= \$36\,000$ A1   **C2**

b Value at end of year 3 = $36\,000 \times (0.88)^2$ M1

$= \$27\,878.40$ A1ft   **C2**

c $\$45\,000 \div 3 = \$15\,000$

If $36\,000 \times (0.88)^n = 15\,000$ then $n \approx 6.85$ {technology}   M1

$\therefore$ the number of years is $7 + 1 = 8$ A1ft   **C2**

[6 marks]

14 a $\qquad y = (x-2)^2 - 5$

$\therefore \quad y = x^2 - 4x + 4 - 5$

$\therefore \quad y = x^2 - 4x - 1$ A1

So, $\dfrac{dy}{dx} = 2x - 4$ A1 A1ft   **C3**

b At $(3, -4)$, $\therefore \dfrac{dy}{dx} = 2 \times 3 - 4 = 2$

So, the gradient of the tangent is 2.   A1ft   **C1**

c The equation of the tangent is $\quad y - -4 = 2(x - 3)$ M1

$\therefore \quad y + 4 = 2x - 6$

$\therefore \quad y = 2x - 10$ A1ft   **C2**

[6 marks]

15 a The sum of the interior angles $= 3 \times 180° = 540°$ M1

So, each interior angle is $\dfrac{540°}{5} = 108°$ A1ft   **C2**

b Area of $\triangle$EAB $= \frac{1}{2} \times 15 \times 15 \times \sin 108°$ M1

$\approx 106.994 \text{ cm}^2$ A1ft

$\therefore$ total area $= (2 \times 106.994 + 173.12) \text{ cm}^2$ M1

$\approx 387.11 \text{ cm}^2$

$\approx 387 \text{ cm}^2$ A1ft   **C4**

[6 marks]

PAPER 2

1 a **i** Number of employees $= 25 \times 1.04 = 26$ A1

     ii Number of employees $= 25 \times (1.04)^{15} \approx 45$ M1 A1 (G2)   **3**

b At the start of 1996, the number of employees

$= 25 \times (1.04)^6$ M1

≈ 32 AG   **1**

c **i** $a = (1.045)^{15} \approx 1.935$ A1

     For $b\%$ interest, $\left(1 + \frac{b}{100}\right)^{10} = 1.877$ M1

     which has solution $b \approx 6.5$ {using technology}   A1 (G2)   **3**

     ii The average salary in 2005 $= 18\,000 \times 2.232 = 40\,176$ NZD   M1 A1 (G2)   **2**

     iii We need to solve for n when $\quad 18\,000 \times (1.045)^n = 30\,000$ M1

         Using technology, $n \approx 11.6$ A1

     So, this occurs in the 12th year, that is, in 2001.   A1 (G2)   **3**

[12 marks]

2 a i $n = 15$ $\therefore$ median = 8th score = 73

   ii Q_1 = 4th score = 68

   Q_3 = 12th score = 79

   $\therefore$ IQR = $Q_3 - Q_1$ = 11

A1ft

M1 A1ft (G2)   **3**

b

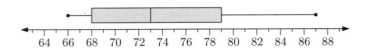

A1 A1 A1   **3**

c mean = $\dfrac{\text{sum of scores}}{15}$

   $= \dfrac{1114}{15}$

   $\approx 74.266\,67$

   ≈ 74.3

M1

A1ft (G2)   **2**

d

Score	60 - 69	70 - 79	80 - 89
Frequency	4	8	3

A1ft A1ft   **2**

e Midpoints are 64.5, 74.5, 84.5

   $\therefore$ an estimate of the mean is $= \dfrac{4 \times 64.5 + 8 \times 74.5 + 3 \times 84.5}{15}$

   $= \dfrac{1107.5}{15}$

   $\approx 73.833\,33$

   ≈ 73.8

M1

A1ft (G2)   **2**

f % error $= \left| \dfrac{74.266\,67 - 73.833\,33}{74.266\,67} \right| \times 100\% \approx 0.583\%$

M1 A1ft (G2)   **2**

[14 marks]

3 a

Hour	Frequency	Cumulative frequency
7:00 am to 8:00 am	14	14
8:00 am to 9:00 am	48	62
9:00 am to 10:00 am	35	97
10:00 am to 11:00 am	24	121
11:00 am to 12 noon	22	143

A1ft A1ft   **2**

b

Traffic in Holden Street

A1
A1 A1ft
A1   **4**

c i median $= \left(\dfrac{143 + 1}{2} \right)$ th score

   = 72nd score

   $\approx$ 9:15 am

A1ft

   ii The 100th vehicle drove down the street at about 10:10 am.

A1ft

   iii The cumulative frequency for 9:30 am is 80, and for 11:30 am is 132

   $\therefore$ the number of cars $\approx 132 - 80 \approx 52$

A1ft

iv 25th percentile $= \dfrac{143 + 1}{4} = $ 36th score which is $\approx$ 8:30 am   A1ft   **4**

[10 marks]

4 a

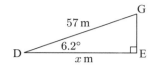

In $\triangle$DGE, $\cos 6.2° = \dfrac{x}{57}$

$\therefore$ $x = 57 \times \cos 6.2°$ M1

$\therefore$ $x \approx 56.67$ A1

So, EF $= 200 - x$

$\approx 200 - 56.67$

≈ 143.33 m

The poles are 143 m apart.   A1ft   **3**

b

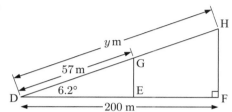

In $\triangle$DHF, $\cos 6.2° = \dfrac{200}{y}$

$\therefore$ $y = \dfrac{200}{\cos 6.2°} \approx 201.18$ M1 A1

$\therefore$ GH $\approx 201.18 - 57$

≈ 144.18

≈ 144 m   A1ft   **3**

c

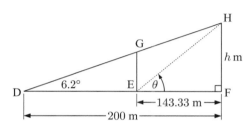

First we need to find FH.

$\tan 6.2° = \dfrac{h}{200}$

$\therefore$ $h = 200 \times \tan 6.2°$ M1

$\therefore$ $h \approx 21.73$ A1

$\tan \theta \approx \dfrac{21.73}{143.33}$ M1

$\therefore$ $\theta \approx 8.62°$

$\therefore$ $\widehat{\text{FEH}} \approx 8.62°$ A1ft   **4**

d

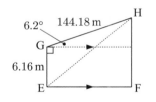

$\sin 6.2° = \dfrac{h}{57}$

$\therefore$ $h = 57 \times \sin 6.2°$

≈ 6.16 m   M1 A1   **2**

e

$\widehat{\text{EGH}} = 90° + 6.2°$

$= 96.2°$ A1

Area of $\triangle$EGH $\approx \frac{1}{2} \times 6.16 \times 144.18 \times \sin 96.2°$ M1 A1ft

≈ 441 m^2 A1ft   **4**

[16 marks]

5 a i

p	q	r	$(p \wedge q)$	$\neg r$	$(p \wedge q) \Rightarrow r$	$(\neg p \vee \neg q)$	$(\neg p \vee \neg q) \Rightarrow \neg r$
T	T	T	T	F	T	F	T
T	T	F	T	T	F	F	T
T	F	T	F	F	T	T	F
T	F	F	F	T	T	T	T
F	T	T	F	F	T	T	F
F	T	F	F	T	T	T	T
F	F	T	F	F	T	T	F
F	F	F	F	T	T	T	T

A1 A1 A1ft   **3**

ii None of these.   A1ft   **1**

iii If the weather is not fine, or the bus is not late, then I will not walk to school.   A1 A1   **2**

iv If the bus is not late then I will walk to school.   A1 A1   **2**

b i P(fine then fine) $= 0.7 \times 0.7$

$= 0.49$ M1 A1 (G2)   **2**

Mathematical Studies SL – Exam Preparation & Practice Guide (3rd edition)

ii	P(the bus is not late) = 0.6	A1
	P(fine and not late) = 0.7×0.6	
	$= 0.42$	A1
	So, P(fine and not late **twice**) = 0.42×0.42	M1
	≈ 0.176	A1ft(G2) **4**

[14 marks]

6 a

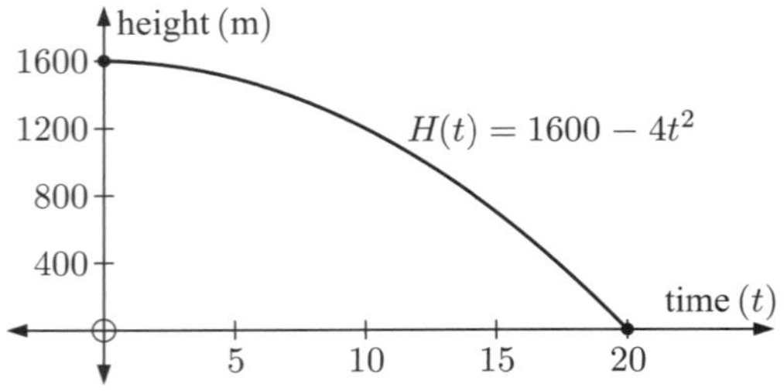

A1 A1 A1   **3**

b i When $H(t) = 0$, $1600 - 4t^2 = 0$ M1

$$\therefore \quad 4t^2 = 1600$$
$$\therefore \quad t^2 = 400$$
$$\therefore \quad t = 20 \ \{t \geqslant 0\}$$ M1

$\therefore$ the object takes 20 seconds to hit the ground.   AG   **2**

ii When $t = 0$, $H(0) = 1600$ m   M1

$\therefore$ vertical height $= 1600$ m   A1(G2)   **2**

iii Average speed $= \dfrac{\text{total distance fallen}}{\text{total time taken}}$

$$= \frac{1600 \text{ m}}{20 \text{ sec}}$$ M1

$$= 80 \text{ ms}^{-1}$$ A1(G2)   **2**

c i $H(10) = 1600 - 4 \times 10^2$ M1

$$= 1200 \text{ m}$$ A1(G2)   **2**

ii $H(10 + h) = 1600 - 4(10 + h)^2$ M1

$$= 1600 - 4(100 + 20h + h^2)$$ M1
$$= 1600 - 400 - 80h - 4h^2$$ M1
$$= 1200 - 80h - 4h^2$$ AG   **3**

iii $\dfrac{H(10 + h) - H(10)}{h} = \dfrac{1200 - 80h - 4h^2 - 1200}{h}$ M1

$$= \frac{-80h - 4h^2}{h}$$

$$= \frac{-80h}{h} - \frac{4h^2}{h}$$

$$= -80 - 4h$$ A1   **2**

d i $H'(t) = 0 - 8t = -8t \text{ ms}^{-1}$ A1 A1   **2**

ii For $t > 0$, $-8t$ is negative.

$\therefore$ $H'(t)$ is negative for all $t > 0$
$\therefore$ H is a decreasing function.   A1 R1   **2**

iii $H'(5) = -8 \times 5 = -40 \text{ ms}^{-1}$ A1

$H'(10) = -8 \times 10 = -80 \text{ ms}^{-1}$ A1   **2**

iv The gradient at $t = 10$ is -80 compared with -40 at $t = 5$

$\therefore$ the graph gets steeper as t increases.   A2   **2**
(This shows that the speed increases as t increases.)

[24 marks]

SOLUTIONS TO TRIAL EXAMINATION 5

PAPER 1

1 a $x^2yz^3 = 5^2 \times 12 \times 100^3$ M1

$$= 300\,000\,000$$ A1

$$= 3 \times 10^8$$ A1ft A1ft   **C4**

b $\sqrt{x^2yz^3} = \sqrt{300\,000\,000}$ M1

            $\approx 17\,320.508\,08$

            $\approx 17\,000$ {to 2 significant figures}   A1ft   **C2**

                                                                    [6 marks]

2   **a**   38, 35, 32, 32, 30, 28, 27, 24, 19, 18   A1   **C1**

    b $n = 10$

        $\therefore$ median $= \dfrac{\text{5th} + \text{6th}}{2} = \dfrac{30 + 28}{2} = 29$ A1ft   **C1**

    c   P(within 5 of median)

       = P(the result is from 24 to 34 inclusive)

       $= \frac{6}{10}$

       $= \frac{3}{5}$ or 0.6   A1 A1ft   **C2**

    d   P(first student is above 30) $= \frac{4}{10}$

       P(2nd student is above 30) $= \frac{3}{9}$

         $\therefore$ P(both above 30) $= \frac{4}{10} \times \frac{3}{9} = \frac{2}{15}$ M1 A1ft   **C2**

                                                                      [6 marks]

3   **a**   {equilateral triangles} is a subset of A,

       so is {isosceles triangles}, and so on.   A1 A1   **C2**

    b   A possible element of B' is a pentagon.   A1   **C1**

    c

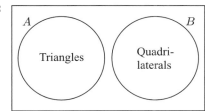

                                                     $U = \{\text{polygons}\}$ A2   **C2**

    d $n(A \cap B) = 0$ A1   **C1**

                                                                        [6 marks]

4   **a**

p	q	$\neg p$	$\neg q$	$(q \wedge \neg p)$	$(\neg q \vee p)$	$(q \wedge \neg p) \Rightarrow (\neg q \vee p)$	
T	T	F	F	F	T	T	
T	F	F	T	F	T	T	A1
F	T	T	F	T	F	F	A1ft
F	F	T	T	F	T	T	A1ft **C3**

    b   Since not all entries are F, the final statement is not a contradiction.   A1ft   **C1**

    c   The contrapositive of the statement is $\neg(\neg q \vee p) \Rightarrow \neg(q \wedge \neg p)$ A1 A1   **C2**

                                                                        [6 marks]

5   **a**   The x-intercepts are -3 and 9.   A1 A1   **C2**

    b $f(x) = 5x^2 + 53x - 84$

       Using technology, $f(x) = 0$ when $x = \frac{7}{5}$ or -12

       $\therefore$ the zeros are $\frac{7}{5}$ and -12.   A1 A1   **C2**

    c   If $y = 3x^2 - 14x + 8$,

       then $\dfrac{dy}{dx} = 6x - 14$ M1 A1   **C2**

                                                                        [6 marks]

6   **a**

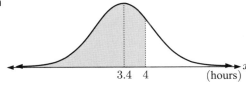

Let X be the time in hours an adult spends watching television on a Saturday night.

Using technology,   P$(X < 4) \approx 0.8043$ A1

The expected number who watch television for less than 4 hours is

         $210 \times 0.8043 \approx 169$ adults   M1 A1   **C3**

Mathematical Studies SL – Exam Preparation & Practice Guide (3rd edition)

b

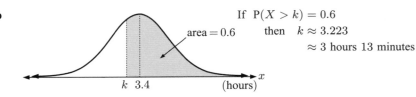

If $P(X > k) = 0.6$ M1
then $k \approx 3.223$ A1
≈ 3 hours 13 minutes   A1   **C3**

[6 marks]

7 a i $y = \frac{3}{2}$ is a horizontal line through $(0, \frac{3}{2})$.

ii For $2x - y - 3 = 0$
when $x = 0$, $y = -3$
when $y = 0$, $x = \frac{3}{2}$

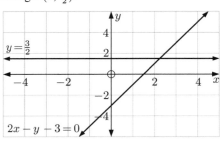

A1 A1 A1   **C3**

b When $y = \frac{3}{2}$, $2x - \frac{3}{2} - 3 = 0$

$$\therefore \quad 2x = \frac{9}{2}$$

$$\therefore \quad x = \frac{9}{4}$$

$$\therefore \quad x = 2\frac{1}{4}$$ M1 A1   **C2**

c The distance is $1\frac{1}{2}$ units.   A1   **C1**

[6 marks]

8 a $\frac{32}{16} = 2$, $\frac{64}{32} = 2$, and so on. So, $r = 2$.   A1   **C1**

b $u_1 = \frac{1}{64}$ and $u_n = 256$

But $u_n = u_1 \times r^{n-1}$

$$\therefore \quad 256 = \frac{1}{64} \times 2^{n-1}$$ M1

$$\therefore \quad 16\,384 = 2^{n-1}$$

$$\therefore \quad n = 15 \quad \{\text{using technology}\}$$

$\therefore$ there are 15 terms in the sequence.   A1ft   **C2**

c $S_n = \dfrac{u_1(r^n - 1)}{r - 1}$

$$= \frac{\frac{1}{64}(2^{15} - 1)}{2 - 1}$$ M1 A1ft

$$= \frac{1}{64}(2^{15} - 1)$$

$$= \frac{32\,767}{64} \quad \{\text{exact answer}\}$$

$$\approx 512$$ A1ft   **C3**

[6 marks]

9 a There are 19 seedlings in the sample.   A1   **C1**

b Since $n = 19$, $\dfrac{n+1}{2} = 10$

i $Q_1 = $ 5th score $= 78$ cm   A1
ii median $= $ 10th score $= 88$ cm   A1
iii $Q_3 = $ 15th score $= 103$ cm   A1   **C3**

c

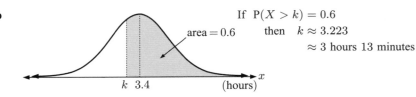

A2ft   **C2**

[6 marks]

10 $C(x) = 46\,000 + 11.50x$ dollars, $R(x) = 30x$ dollars

  a **i** When $x = 2000$,

$$\text{profit} = R - C$$
$$= (30 \times 2000) - (46\,000 + 11.5 \times 2000) \qquad \text{M1}$$
$$= -\$9000$$

This indicates a loss of $9000.   A1   **C2**

  ii When $x = 5000$,

$$\text{profit} = R - C$$
$$= (30 \times 5000) - (46\,000 + 11.5 \times 5000) \qquad \text{M1}$$
$$= \$46\,500 \qquad \text{A1} \qquad \textbf{C2}$$

  b The breakeven point occurs when $C(x) = R(x)$

$$\therefore \quad 30x = 46\,000 + 11.5x \qquad \text{M1}$$
$$\therefore \quad x \approx 2486.486 \quad \{\text{using technology}\}$$

So, the publisher breaks even when 2487 books are sold.   A1   **C2**

[6 marks]

11

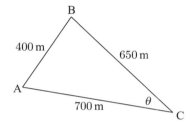

  a Using the cosine rule,

$$\cos\theta = \frac{700^2 + 650^2 - 400^2}{2 \times 700 \times 650} \qquad \text{M1 A1}$$
$$\therefore \quad \cos\theta \approx 0.826\,923$$
$$\therefore \quad \theta \approx 34.216°$$
$$\therefore \quad \widehat{BCA} \approx 34.2° \qquad \text{A1} \qquad \textbf{C3}$$

  b Area $= \frac{1}{2} \times 700 \times 650 \times \sin\theta$ M1 A1ft

$$\approx 127\,927$$
$$\approx 128\,000 \text{ m}^2 \qquad \text{A1ft} \qquad \textbf{C3}$$

[6 marks]

12 **a** **i** The final value $= 3500\left(1 + \frac{5.2}{100}\right)^7$ M1

$$= 4990.89 \text{ pesos} \qquad \text{A1} \qquad \textbf{C2}$$

  ii $A = 3500 \times \left(1 + \frac{5.2}{1200}\right)^{7 \times 12}$ M1 A1

$$\approx 5032.80 \text{ pesos} \qquad \text{A1} \qquad \textbf{C3}$$

  b The difference in interest paid $= (5032.80 - 4990.89)$ pesos

$$= 41.91 \text{ pesos} \qquad \text{A1ft} \qquad \textbf{C1}$$

[6 marks]

13 **a** The x-intercepts are at $(-3, 0)$ and $(4, 0)$.   A1   **C1**

  b The axis of symmetry is midway between the x-intercepts.

$$\frac{-3 + 4}{2} = \frac{1}{2} \quad \therefore \quad \text{the axis of symmetry is } x = \tfrac{1}{2}. \qquad \text{A1ft} \qquad \textbf{C1}$$

  c The function has equation $y = a(x + 3)(x - 4)$, $a \neq 0$ M1

$$\text{But when } x = 0, \quad y = -24$$
$$\therefore \quad -24 = a(3)(-4)$$
$$\therefore \quad a = 2$$
$$\therefore \quad \text{the function is } y = 2(x + 3)(x - 4) \qquad \text{A1ft} \qquad \textbf{C2}$$

  d When $x = \frac{1}{2}$, $y = 2\left(\frac{7}{2}\right)\left(-\frac{7}{2}\right) = -\frac{49}{2}$ M1

Since the graph opens upwards, the minimum value is -24.5 when $x = \frac{1}{2}$.   A1ft   **C2**

[6 marks]

14 a $Q \approx -0.543P + 74.3$ with $r \approx -0.642$ A1 A1   **C2**

b The scatterplot is:

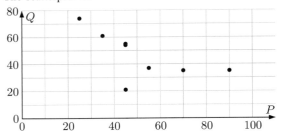

The outlier is the point $(45, 21)$.

After removing the outlier, $Q \approx -0.614P + 82.2$ with $r \approx -0.897$.   A1 A1   **C2**

c The gradient is more negative, so the graph is steeper.   A1ft

As r is closer to -1, the relationship is much stronger.   A1ft   **C2**

[6 marks]

15 a i $f'(x) > 0$ when the function is increasing.

This is when $x < b$ and when $x > c$.   A1 A1   **C2**

ii $f'(x) < 0$ when the function is decreasing.

This is when $b < x < c$.   A1ft A1ft   **C2**

b $f'(x) = 0$ when the tangents are horizontal. This is at the local maximum and minimum points, when $x = b$ and when $x = c$.   A1 A1   **C2**

[6 marks]

PAPER 2

1 a i Total rent $= 700 \times 12 \times 3$

$= 25\,200$ USD   A1   **1**

ii Total rent increase at end of year $1 = 25\,200 \times 0.028$

$= 705.60$ USD   M1 A1ft (G2)   **2**

iii In the third year each person pays $700 \times 12 \times 1.028 \times 1.033 \approx 8920.16$ USD   M1 M1 A1ft (G3)   **3**

b i $u_1 = 800,\quad d = 40$

$u_5 = u_1 + 4d = 800 + 160 = 960$ M1

At the start of 2009 there were 960 employees.   A1 (G2)   **2**

ii $u_1 - 5d = 800 - 200 = 600$ M1

At the start of 2000 there were 600 employees.   A1 (G2)   **2**

iii $u_n = 800 + (n - 1) \times 40$

The number of employees is greater than 1000 when $800 + (n - 1) \times 40 > 1000$ M1

$\therefore\ n > 6$ A1

So, there will be more than 1000 employees in 2011.   A1 (G2)   **3**

[13 marks]

2 a i Length $= (x + 5)$ cm   A1   **1**

ii

x cm

$(x+5)$ cm

Area $=$ length $\times$ width

$\therefore\ A = (x + 5)x$

$= (x^2 + 5x)$ cm^2 A1 A1ft   **2**

iii The area is 204 cm^2 when $x^2 + 5x = 204$ M1

$\therefore\ x^2 + 5x - 204 = 0$

This has solutions $x = 12$ or -17.   A1ft (G2)

Since lengths must be non-negative, $x = 12$.   A1ft   **3**

iv So, the length is 17 cm and the width is 12 cm.   A1ft   **1**

b i The volume of a sphere $= \frac{4}{3}\pi r^3$

$\therefore\ \frac{1}{2}\left(\frac{4}{3}\pi r^3\right) = 7068$ m^3 M1 A1

$\therefore\ \frac{2}{3}\pi r^3 = 7068$

$\therefore\ r^3 = \dfrac{7068 \times 3}{2 \times \pi}$

$\therefore\ r \approx 15.0$ m {using technology}   A1 (G2)   **3**

ii The surface area of a sphere $= 4\pi r^2$

∴ the surface area of an open hemisphere $= 2\pi r^2$

$$\approx 2 \times \pi \times (15.0)^2 \qquad \text{M1 A1ft}$$
$$\approx 1413.72$$
$$\approx 1410 \text{ m}^2 \qquad \text{A1ft(G1)} \qquad 3$$

iii There are 142 lots of 10 m^2 required.

∴ the total cost $= 142 \times \$23.45 = \$3329.90$ \qquad M1 A1ft(G2) \qquad 2

c i

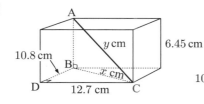

In $\triangle$BCD, $\quad x^2 = 10.8^2 + 12.7^2$ \qquad M1

∴ $x \approx 16.67$ \qquad A1

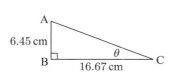

In $\triangle$ABC, $\quad y^2 = x^2 + 6.45^2$

So, $\quad y^2 = 10.8^2 + 12.7^2 + 6.45^2$ \qquad M1

∴ $\quad y = \sqrt{10.8^2 + 12.7^2 + 6.45^2}$

≈ 17.8755

∴ the longest piece of string is 17.9 cm. \qquad A1ft(G2) \qquad 4

ii

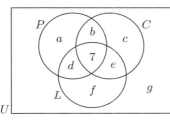

$\tan \theta = \dfrac{6.45}{16.67}$ \qquad M1

∴ $\theta = \tan^{-1}\left(\dfrac{6.45}{16.67}\right)$

∴ $\theta \approx 21.2°$

∴ the string makes an angle of about 21.2° with the base of the box. \qquad A1ft(G1) \qquad 2

[21 marks]

3 a i

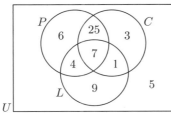

$n(U) = 60$

$g = 5 \quad$ {5 have none of P, C or L}

$a = 6 \quad$ {6 contain peppermints only}

$7 + e = 8 \quad$ {8 contain some of each of L and C}

∴ $e = 1$

Likewise $d = 4 \quad$ {11 contain some of each of P and L}

So, $b + 6 + 4 + 7 = 42 \qquad ∴ \quad b = 25$

$c + 25 + 7 + 1 = 36 \qquad ∴ \quad c = 3$

$f + 4 + 7 + 1 = 21 \qquad ∴ \quad f = 9$

$n(U) = 60$

\qquad A1 A1 A1 \qquad 3

ii (1) $n(P \cap C) = 25 + 7 = 32$ \qquad A1ft

(2) $n(P \cup C') = 60 - 1 - 3 = 56$ \qquad A1ft

(3) $n\big((L \cap P)' \cap (C \cup L)\big) = 9 + 1 + 3 + 25$

$= 38$ \qquad A1ft A1ft \qquad 4

b i Total number $= 15 + 8 + 12 + ... + 14$

$= 110$ items \qquad A1 \qquad 1

ii (1) P(gold earrings) $= \frac{10}{110} = \frac{1}{11}$ \qquad A1 A1 (G2) \qquad 1

(2) P(not a silver bracelet) $= \frac{110-12}{110} = \frac{98}{110} = \frac{49}{55}$ \qquad A1 A1 (G2) \qquad 2

iii   P(both silver | both necklaces)

$$= \frac{8}{42} \times \frac{7}{41}$$ M1

$$= \frac{4}{123}$$ A1 A1 (G2)   **3**

[15 marks]

4   **a**   **i** $a = f(-1)$

$\qquad = 2(-1)^2 - 6(-1) - 20$

$\qquad = -12$ A1

$\qquad b = f(2)$

$\qquad = 2(2)^2 - 6(2) - 20$

$\qquad = -24$ A1   **2**

ii $f(-3) = 2(-3)^2 - 6(-3) - 20$ and $f(7) = 2(7)^2 - 6(7) - 20$

$\qquad\qquad = 18 + 18 - 20 \qquad\qquad\qquad\quad = 98 - 42 - 20$

$\qquad\qquad = 16 \qquad\qquad\qquad\qquad\qquad\quad = 36$

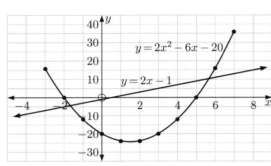

                 Diagram reduced 50%   A1 A2ft

                                                 A1ft   **4**

iii   see graph above   A2   **2**

iv   They meet where $2x^2 - 6x - 20 = 2x - 1$

$\qquad\qquad\qquad \therefore \quad 2x^2 - 8x - 19 = 0$

$\qquad\qquad\qquad\qquad \therefore \quad x \approx -1.67 \text{ or } 5.67 \ \{\text{using technology}\}$

Points of intersection are $(-1.67, -4.35)$ and $(5.67, 10.4)$.   A1 A1   **2**

b $f(x) = \frac{5}{3}x^3 + 3x^2 - 8x + 2, \quad 0 \leqslant x \leqslant 3$

   i $f'(x) = 5x^2 + 6x - 8$ A1 A1 A1   **3**

   ii $f'(1) = 5 + 6 - 8 = 3$

$\qquad \therefore \quad$ the gradient of the normal at $x = 1$ is $-\frac{1}{3}$.   M1 A1   **2**

   iii $f'(x) = 0$ when $5x^2 + 6x - 8 = 0$ M1

$\qquad \therefore \quad (5x - 4)(x + 2) = 0$

$\qquad\qquad \therefore \quad x = \frac{4}{5} \text{ or } -2$ A1ft A1ft (G3)   **3**

   iv $f(0.8) \approx -1.63$ and $f(-2) \approx 16.67$

But $x = -2$ is outside of the domain $0 \leqslant x \leqslant 3$.   A1ft

We sketch $y = f(x)$ using technology:

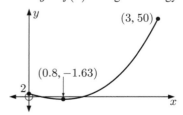

$\qquad \therefore \quad f(x)$ has largest value 50 when $x = 3$ A2

and smallest value -1.63 when $x = 0.8$ A1 A1ft   **5**

[23 marks]

5 a

Height	Frequency	Midpoint
0 - 4	0	2
5 - 9	0	7
10 - 14	2	12
15 - 19	8	17
20 - 24	20	22
25 - 29	30	27
30 - 34	24	32
35 - 39	16	37
Total	100	

A1 A1   **2**

b mean ≈ 27.7,   standard deviation ≈ 6.25 A1ft A1ft   **2**

c The % difference in means $\approx \dfrac{27.7 - 23}{27.7} \times 100\%$ M1

$\approx 17.0\%$ A1ft (G1)   **2**

d The height 0.368 standard deviations above the mean $\approx 27.7 + 0.368 \times 6.25$

≈ 30 cm   M1

From the graph, 64 plants have height $\leqslant 30$ cm

$\therefore$ 36 have height > 30 cm   A1ft (G2)   **2**

[8 marks]

6 a H_0: Productivity and training are independent.   A1

H_1: Productivity and training are not independent.   A1   **2**

b $\chi^2_{calc} = 1.67$ G2   **2**

c The number of degrees of freedom is $(3-1)(2-1) = 2$ A1 (AG)   **1**

d Since $\chi^2_{calc} < \chi^2_{crit}$, we accept the null hypothesis and conclude that at the
5% level of significance, productivity and training are independent.   A1ft R1ft   **2**

e For the new data, $\chi^2_{calc} = 7.28$ G1

In this case, $\chi^2_{calc} > \chi^2_{crit}$, so we reject the null hypothesis and conclude that
at the 5% level of significance, productivity and training are not independent.   A1 R1   **3**

[10 marks]

NOTES

NOTES

Mathematical Studies SL – Exam Preparation & Practice Guide (3rd edition)

NOTES